# MATHS PLUS

## MENTALS AND HOMEWORK BOOK

AUSTRALIAN CURRICULUM

Harry O'Brien
Greg Purcell

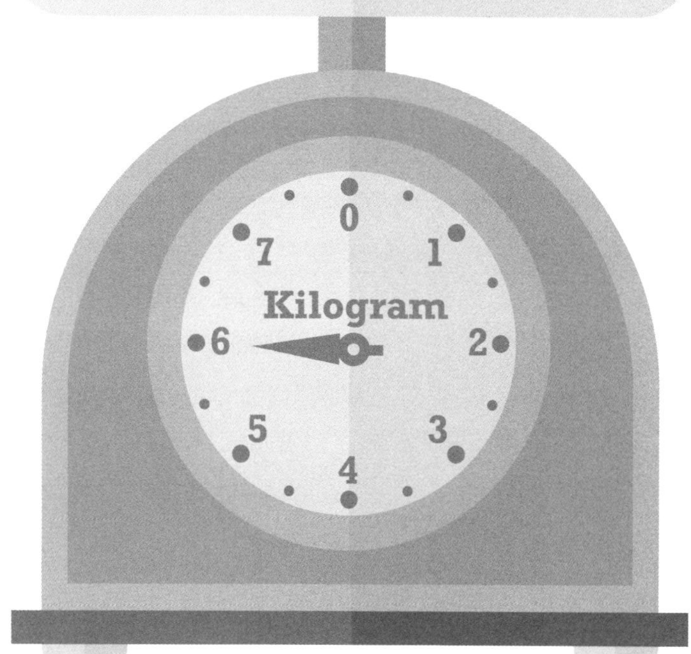

OXFORD
UNIVERSITY PRESS

# Contents

# Contents

# Australian Curriculum Linking Chart

| Units | 1 | 2 | 3 | 4 | 5 | 6 |
|---|---|---|---|---|---|---|
| **Number and Algebra** | | | | | | |
| Recognise and extend the application of place value to tenths and hundredths and use the conventions of decimal notation to name and represent decimals (AC9M4N01) | | | | | | |
| Explain and use the properties of odd and even numbers (AC9M4N02) | | | | | | |
| Find equivalent representations of fractions using related denominators and make connections between fractions and decimal notation (AC9M4N03) | | | | | | |
| Count by fractions including mixed numerals; locate and represent these fractions as numbers on number lines (AC9M4N04) | | | | | | |
| Solve problems involving multiplying or dividing natural numbers by multiples and powers of 10 without a calculator, using the multiplicative relationship between the place value of digits (AC9M4N05) | | | | | | |
| Develop efficient strategies and use appropriate digital tools for solving problems involving addition and subtraction, and multiplication and division where there is no remainder (AC9M4N06) | | | | | | |
| Choose and use estimation and rounding to check and explain the reasonableness of calculations including the results of financial transactions (AC9M4N07) | | | | | | |
| Use mathematical modelling to solve practical problems involving additive and multiplicative situations including financial contexts; formulate the problems using number sentences and choose efficient calculation strategies, using digital tools where appropriate; interpret and communicate solutions in terms of the situation (AC9M4N08) | | | | | | |
| Follow and create algorithms involving a sequence of steps and decisions that use addition or multiplication to generate sets of numbers; identify and describe any emerging patterns (AC9M4N09) | | | | | | |
| Find unknown values in numerical equations involving addition and subtraction, using the properties of numbers and operations (AC9M4A01) | | | | | | |
| Recall and demonstrate proficiency with multiplication facts up to 10 x 10 and related division facts; extend and apply facts to develop efficient mental strategies for computation with larger numbers without a calculator (AC9M4A02) | | | | | | |
| **Measurement and Space** | | | | | | |
| Interpret unmarked and partial units when measuring and comparing attributes of length, mass, capacity, duration and temperature, using scaled and digital instruments and appropriate units (AC9M4M01) | | | | | | |
| Recognise ways of measuring and approximating the perimeter and area of shapes and enclosed spaces, using appropriate formal and informal units (AC9M4M02) | | | | | | |
| Solve problems involving the duration of time including situations involving "am" and "pm" and conversions between units of time (AC9M4M03) | | | | | | |
| Estimate and compare angles using angle names including acute, obtuse, straight angle, reflex and revolution, and recognise their relationship to a right angle (AC9M4M04) | | | | | | |
| Represent and approximate composite shapes and objects in the environment, using combinations of familiar shapes and objects (AC9M4SP01) | | | | | | |
| Create and interpret grid reference systems using grid references and directions to locate and describe positions and pathways (AC9M4SP02) | | | | | | |
| Recognise line and rotational symmetry of shapes and create symmetrical patterns and pictures, using dynamic geometric software where appropriate (AC9M4SP03) | | | | | | |
| **Statistics and Probability** | | | | | | |
| Acquire data for categorical and discrete numerical variables to address a question of interest or purpose, using digital tools; represent data using many-to-one pictographs, column graphs and other displays or visualisations; interpret and discuss the information that has been created (AC9M4ST01) | | | | | | |
| Analyse the effectiveness of different displays or visualisations in illustrating and comparing data distributions, then discuss the shape of distributions and the variation in the data (AC9M4ST02) | | | | | | |
| Conduct statistical investigations, collecting data through survey responses and other methods; record and display data using digital tools; interpret the data and communicate the results (AC9M4ST03) | | | | | | |
| Describe possible everyday events and the possible outcomes of chance experiments and order outcomes or events based on their likelihood of occurring; identify independent or dependent events (AC9M4P01) | | | | | | |
| Conduct repeated chance experiments to observe relationships between outcomes; identify and describe the variation in results (AC9M4P02) | | | | | | |

# Units

7 8 9 10 11 12 13 14 15 16 17 18 19 20 21 22 23 24 25 26 27 28 29 30 31 32 33 34 35

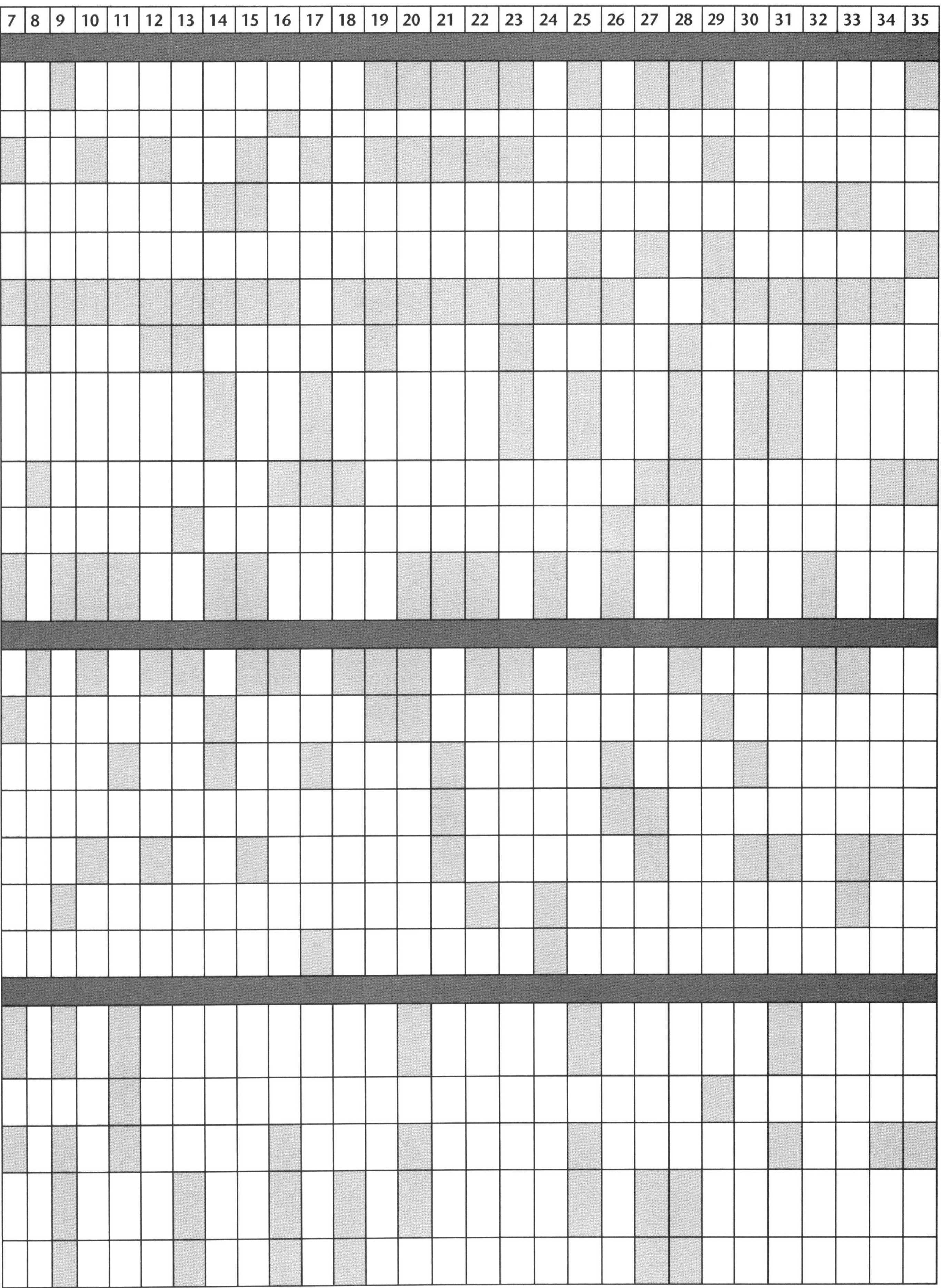

# Number and Algebra

## SET 1 Basic

1 4 + 2 + 6

2 6 + 8 + 3

3 9 – 5

4 20 – 5

5 6 × 5

6 7 × 4

7 2 × 10

8 Which is the third month of the year?

9 What is the sum of 9 and 5?

10 What is the difference between 11 and 8?

11 Subtract 5 from 11.

12 Halve 14c.

13 How many minutes in a quarter of an hour?

14

## SET 2 Split strategy for addition

Solve these additions using the split strategy.

| | Question | Becomes | Answer |
|---|---|---|---|
| 1 | 25 + 46 | 60 + 11 | |
| 2 | 38 + 17 | | |
| 3 | 53 + 24 | | |
| 4 | 68 + 31 | | |
| 5 | 47 + 55 | | |
| 6 | 82 + 26 | | |
| 7 | 133 + 28 | | |
| 8 | 152 + 49 | | |

Use the split strategy to check these additions and circle true or false.

9 62 + 39 = 101 True/False

10 26 + 43 = 79 True/False

11 55 + 37 = 92 True/False

12 112 + 67 = 179 True/False

# Space Three-dimensional revision

Colour code each matching description, name and shape.

1 My 2 bases are circles and my other face is curved.

2 I have 6 faces which are all rectangles.

3 I have 1 base which is a circle.

4

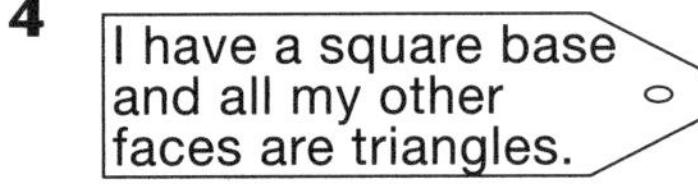


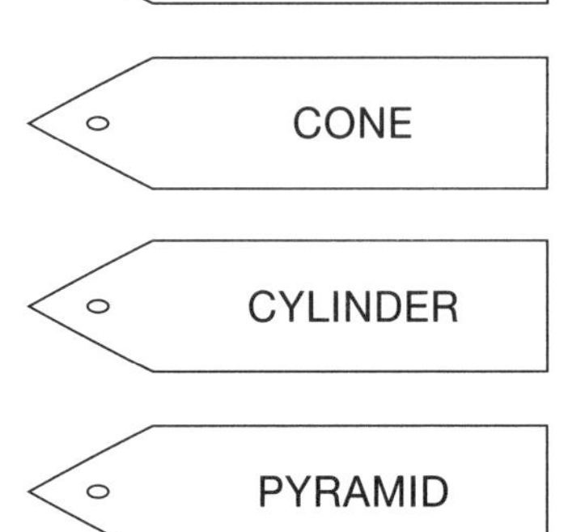


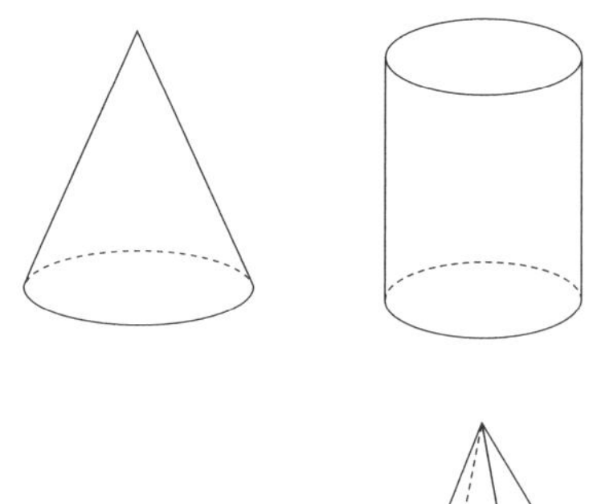

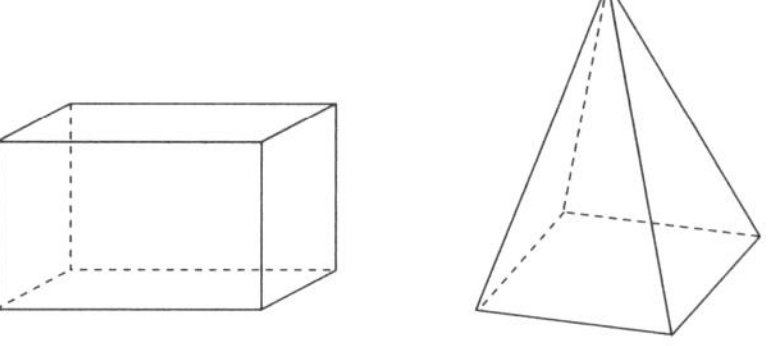

# Number and Algebra

## SET 3 Multiplication facts (3s)

Answer the multiplications.

| | | | | | |
|---|---|---|---|---|---|
| 1 | 2 × 3 = | | 6 | 4 × 3 = | |
| 2 | 3 × 3 = | | 7 | 6 × 3 = | |
| 3 | 1 × 3 = | | 8 | 8 × 3 = | |
| 4 | 5 × 3 = | | 9 | 9 × 3 = | |
| 5 | 10 × 3 = | | 10 | 7 × 3 = | |

Complete the multiplication targets.

11

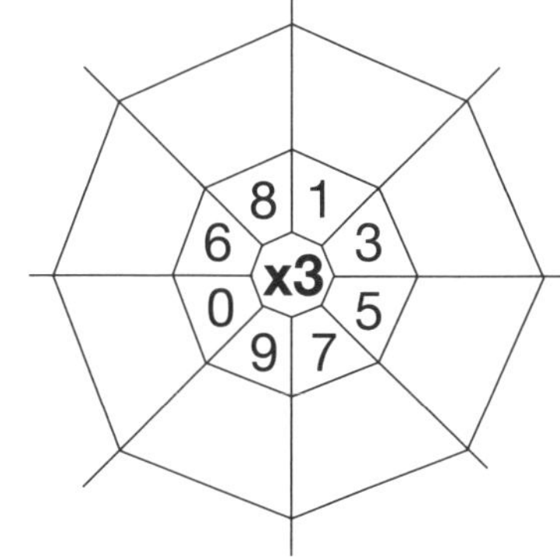


12 Joe had 8 marbles but Lisa had twice that number. How many marbles did Lisa have?

## SET 4 Extension

1 3 dozen take away 17

2 What is the difference between 42 and 8?

3 If today is Wednesday, what day is 11 days later?

4 50, 45, ☐, ☐, ☐, 25

5 6 × ☐ = 54

6 $5.63 = ☐ cents

7 What time is 4 hours after 3:45 pm?

8 What is left from $2.00 if I spent 85c?

9 How much are 8 pens at 90c each?

10 How much are 3 yo-yos at 25c each?

11 Make the largest number possible using 6, 4, 9, 8.

**Mathematical Reasoning**

12 3 girls each paid $75 for their excursion. Name 3 sets of notes that equal $75.

a $ ☐ $ ☐ $ ☐ $ ☐ $ ☐

b $ ☐ $ ☐ $ ☐ $ ☐

c $ ☐ $ ☐ $ ☐

# Measurement Centimetres

Extend these lines to the lengths listed.

5 cm

7 cm

9 cm

13 cm

$14\frac{1}{2}$ cm

$15\frac{1}{2}$ cm

UNIT 2

# Number and Algebra

## SET 1 Basic

1 3 + 4 + 2

2 4 + 6 + 0

3 9 – 6

4 7 – 5

5 3 × 5

6 6 × 4

7 5 × 4

8 5 × 6

9 Which is the first month of the year?

10 What is the sum of 6 and 3?

11 What is the difference between 9 and 4?

12 Add 6 and 1.

13 How many minutes in 1 hour?

14 4 tens + 6 ones

15

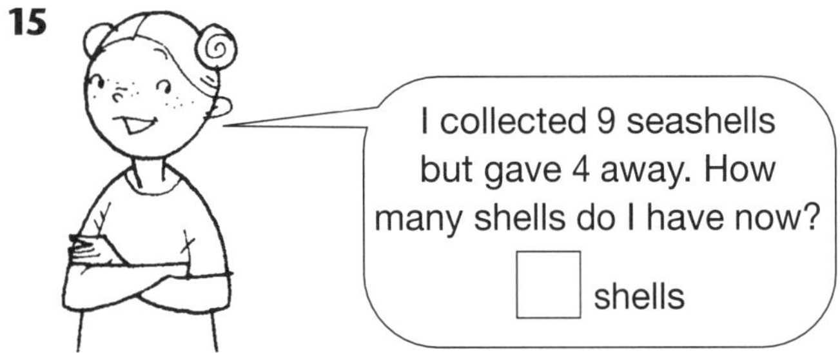


## SET 2 Subtraction strategies

53 – 28 = ?

Think

53 – 20 – 8 = 25

Solve the subtractions.

1 43 – 21 = ☐ 2 65 – 39 = ☐

3 72 – 54 = ☐ 4 89 – 44 = ☐

5 32 – 17 = ☐ 6 58 – 37 = ☐

7 91 – 76 = ☐ 8 64 – 51 = ☐

9 88 – 55 = ☐ 10 75 – 28 = ☐

Extend these subtraction facts.

11 9 – 7 = 14 12 – 5 =

12 90 – 70 = 15 120 – 50 =

13 900 – 700 = 16 1200 – 500 =

Complete the grids.

17

| ● | 86 | 74 | 49 | 24 | 13 | |
|---|---|---|---|---|---|---|
| ▲ | 77 | | 40 | | | 36 |

18

| ● | 35 | 55 | 85 | 95 | 125 | 135 |
|---|---|---|---|---|---|---|
| ▲ | 29 | | 79 | | 119 | |

# Statistics and Probability Column graphs

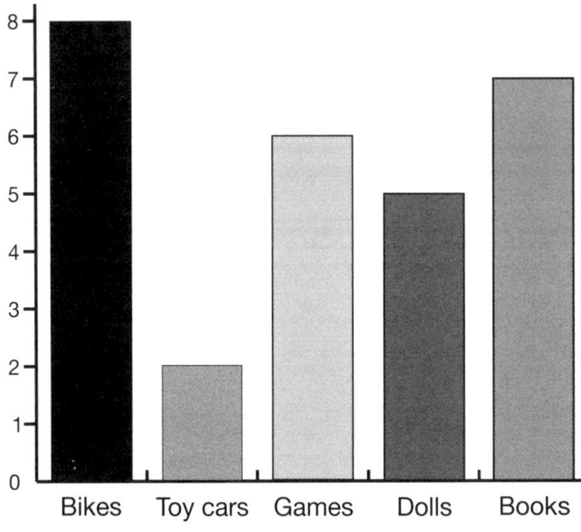


**My favourite thing**

1 How many children like books best? ☐

2 How many children like bikes best? ☐

3 Which is the least popular thing? ☐

4 Which item was chosen by 7 children? ☐

5 How many children were surveyed? ☐

6 Which 2 items together equal bikes?

☐

# Number and Algebra

## SET 3 Halves, quarters and eighths

Shade the fractions.

1 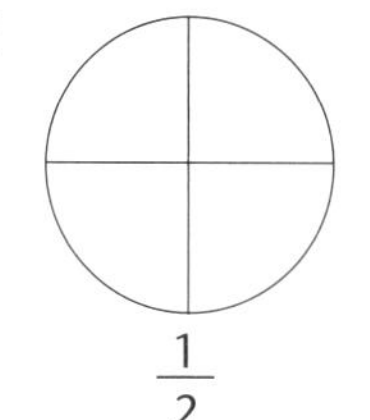 $\frac{1}{2}$

2 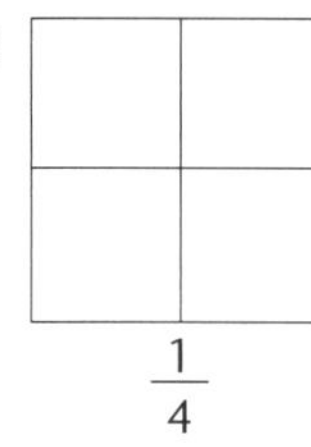 $\frac{1}{4}$

3 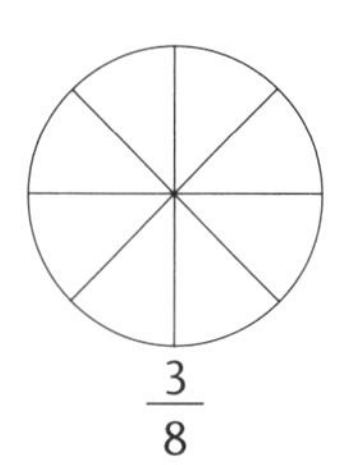 $\frac{3}{8}$

4 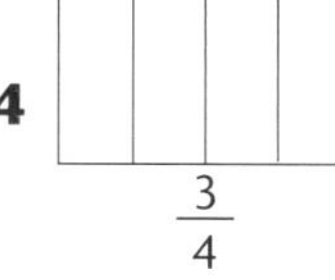 $\frac{3}{4}$

5 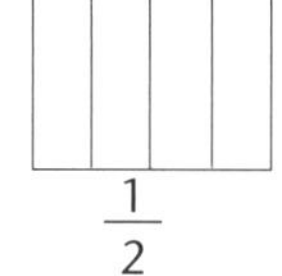 $\frac{1}{2}$

6 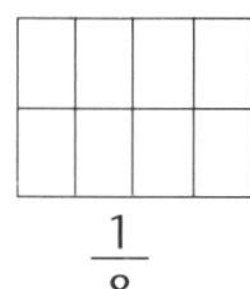 $\frac{1}{8}$

7 □ □ □ □ $\frac{1}{4}$

8 □ □ □ □ □ □ □ □ $\frac{7}{8}$

9 How much fuel is in each tank?

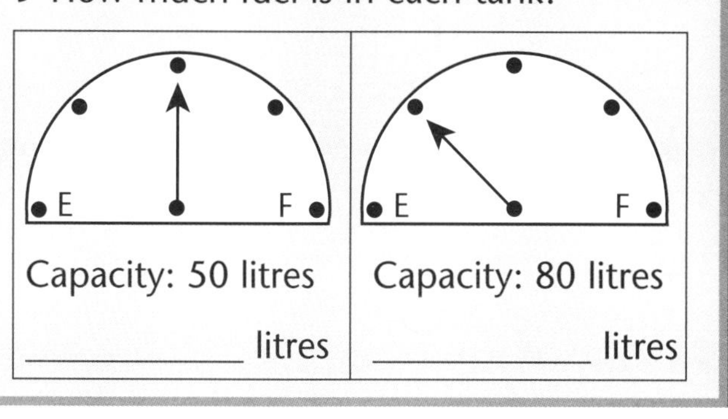


Capacity: 50 litres ________ litres

Capacity: 80 litres ________ litres

Mathematical Reasoning

## SET 4 Extension

1 What is the 3rd letter of the alphabet?

2 Share 50c equally among 5 children.

3 12 + 3 + 4 + 5

4 (3 × 2) + (5 × 2)

5 1786c = $ ☐

6 What is the value of 6 in 7642?

7 Write the numeral for seven hundred and twenty-six.

8 1 fortnight + 8 days

9 If 2 kg cost $6, how much would 5 kg cost?

10 How many faces has a cube?

11 $12 – $1.50

12 $9.87 = ☐ c

13 How many hours from 3 pm to 10 pm?

14 How many legs on 4 dogs and 4 spiders?

15 Write 7260 in words. ________________

________________

16 How many quarters in one half?

# Measurement Estimating length

Estimate and measure the length of these pencils.

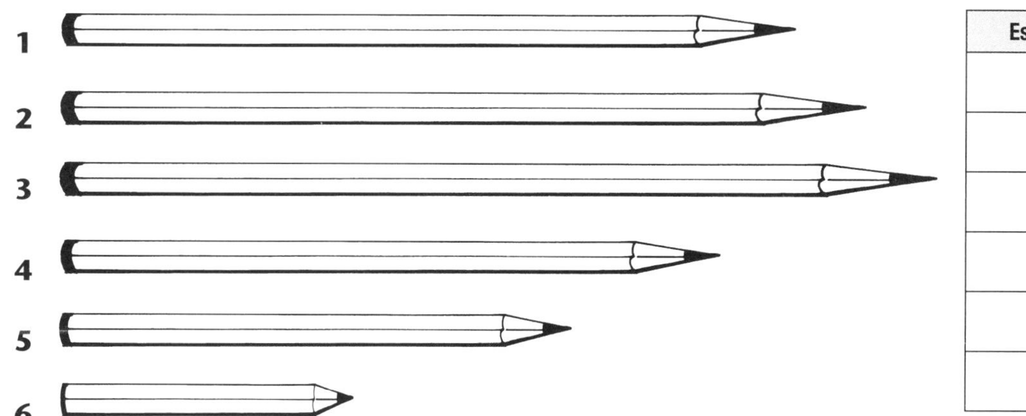

1

2

3

4

5

6

| Estimate | cm |
|---|---|
| | |
| | |
| | |
| | |
| | |
| | |

UNIT 3

# Number and Algebra

## SET 1 Basic

1 5 + 6

2 7 + 3

3 18 – 4

4 25 – 5

5 3 × 6

6 4 × 2

7 5 × 4

8 6 × 3

9 What is the 6th month of the year?

10 What is the sum of 12 and 8?

11 What is the difference between 12 and 8?

12 Triple 2.

13 How many days in a fortnight?

14 How many minutes in 3 hours?

15

## SET 2 Multiplication facts

Complete the table.

| | × | 3 | 2 | 5 | 1 | 0 | 7 | 9 |
|---|---|---|---|---|---|---|---|---|
| 1 | 3 | | | | | | | |
| 2 | 0 | | | | | | | |
| 3 | 4 | | | | | | | |
| 4 | 10 | | | | | | | |

5 \$9 × 5

6 \$3 × 6

7 5 kg of beans at \$4 per kilo

8 6 kg of potatoes at \$5 per kilo

Mathematical Reasoning

9

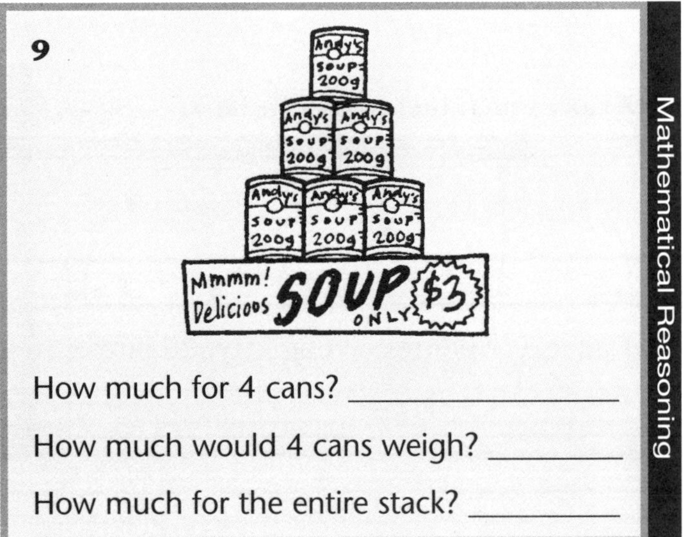


How much for 4 cans? ____________

How much would 4 cans weigh? ________

How much for the entire stack? ________

## Space Combining 2D shapes

How many letters of the alphabet can you make using 5 squares? Use the grid paper to display your answer.

# Number and Algebra

## SET 3 Turnarounds

Complete the sets of additions.

| | | |
|---|---|---|
| 1 | 3 + 7 = | 7 + 3 = |
| 2 | 9 + 8 = | 8 + 9 = |
| 3 | 20 + 70 = | 70 + 20 = |
| 4 | 80 + 70 = | 70 + 80 = |
| 5 | 126 + 6 = | 6 + 126 = |
| 6 | 250 + 26 = | 26 + 250 = |

7 Did the order of adding the numbers change the answer?

Complete the sets of multiplications.

| | | |
|---|---|---|
| 8 | 6 × 5 = | 5 × 6 = |
| 9 | 7 × 4 = | 4 × 7 = |
| 10 | 8 × 3 = | 3 × 8 = |
| 11 | 8 × 6 = | 6 × 8 = |
| 12 | 7 × 5 = | 5 × 7 = |
| 13 | 8 × 4 = | 4 × 8 = |

14 Did the order of multiplying change the answer?

## SET 4 Extension

1 What is the 5th letter of the alphabet?

2 How many times can $4 be taken away from $20?

3 Share $1.00 among 5 children.

4 18c + 17c + 19c + 13c

5 How many days altogether in January and November?

6 What is the value of 6 in 6321?

7 Estimate an answer to 999 + 496.

8 How much is 3 kg of sugar at $1.50 per kg?

9 Round 3921 to the nearest 1000.

10 Is a cylinder a 2D shape?

11 How many 5c coins make $1?

12 10 books at $17 each

13 How many minutes between 1 pm and 4 pm?

14 Seven lots of five

15 Write 3062 in words. ______________________

______________________________

16 Each player receives 15 cards. How many cards do I need for 3 players?

# Measurement Perimeter

Measure the perimeters of these shapes.

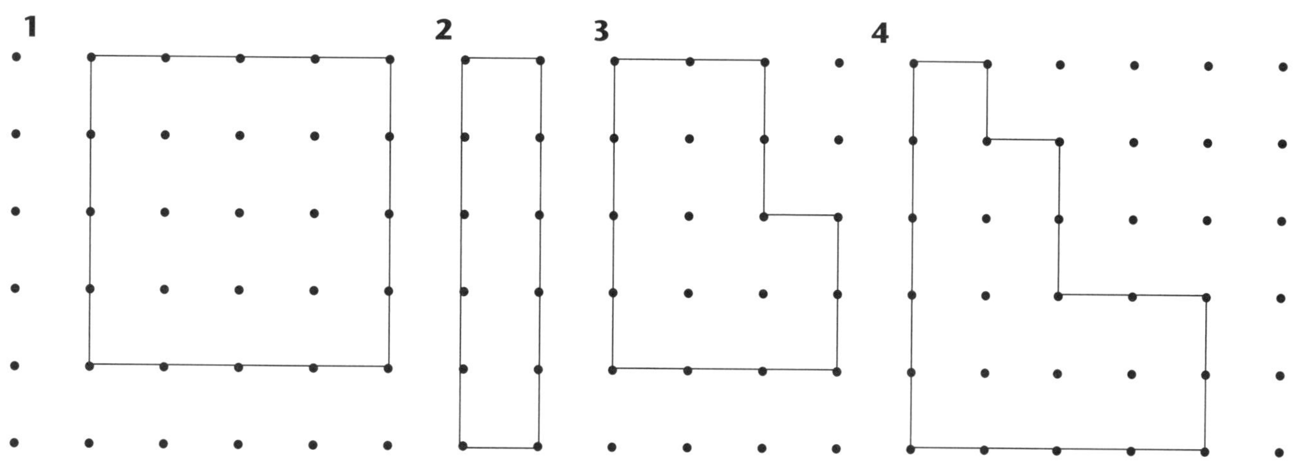

UNIT 4

# Number and Algebra

## SET 1 Basic

1 6 + 4 + 4

2 7 + 3 + 6

3 18 – 5

4 17 – 6

5 4 × 10

6 5 × 9

7 7 × 3

8 6 × 2

9 What is the difference between 43 and 10?

10 Add 24 to 13.

11 Multiply 3 × 6.

12 What is the 4th month of the year? [ ]

13 Minutes in $\frac{1}{2}$ hour

14 3 × 7 + 6

15 How much money have I got if I have one cheque for $3000, two $100 notes, three $10 notes and seven $1 coins?

$ [ ]

## SET 2 Place value to 9999

State the place value of each bold number.

| | Number | Place Value |
|---|---|---|
| 1 | **7**23 | |
| 2 | 8**4**2 | |
| 3 | **3**746 | |
| 4 | 529**6** | |
| 5 | 7**5**26 | |

6 What number is 100 more than 756?

7 What number is 100 more than 2357?

8 What number is 1000 more than 3576?

9 What number is 10 more than 2765?

10 What number is 100 less than 3574?

11 What number is 30 more than 367?

12 What number is 40 more than 1357?

13 What number is 50 less than 3590?

14 What number is 200 more than 3264?

15 1256 + 40

16 Write the largest number you can using 3, 7, 5.

17 Write the smallest number you can using 3, 7, 5.

## Space Grid references

Answer yes or no to the below questions.

1 Does the person at A3 wear glasses?

2 Does the person at C1 have a moustache?

3 Does the person at B1 have a beard?

4 Does the person at C3 have no hair?

5 Does the person at B3 wear glasses?

6 Is the person at B2 wearing earrings?

# Number and Algebra

## SET 3 Fifths and tenths

Write a fraction for each shaded shape.

1 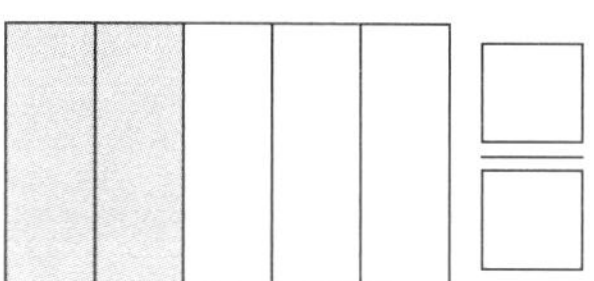

2 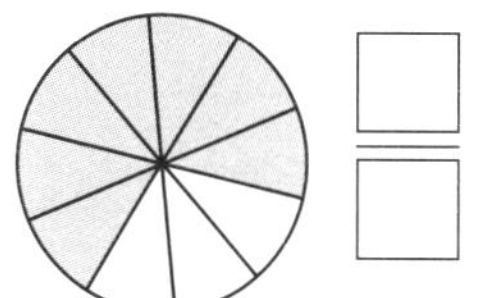

3 

4 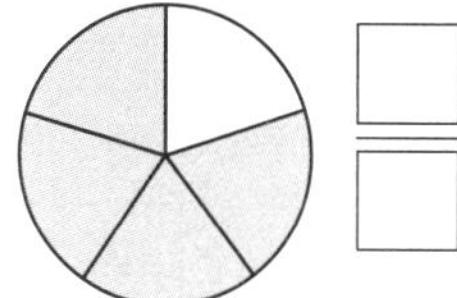

5 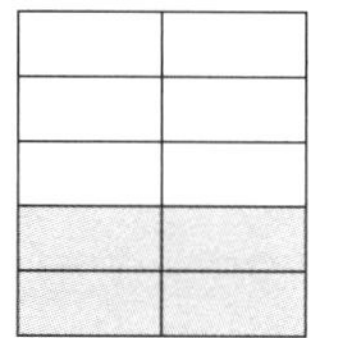

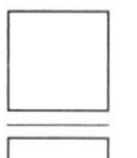

6 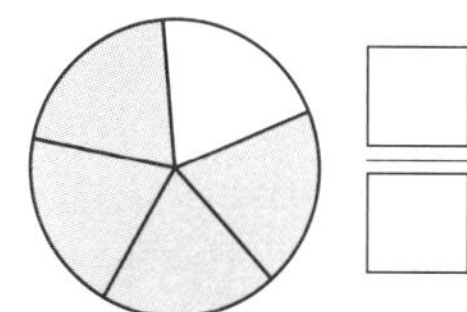

7 Order the fractions from smallest to largest: $\frac{1}{10}$, $\frac{5}{10}$, $\frac{2}{10}$, $\frac{9}{10}$.

8 How many fifths in a whole?

9 Is $\frac{4}{5}$ larger than $\frac{4}{10}$?

10 Does $\frac{3}{5} = \frac{6}{10}$?

11 Write a fraction larger than $\frac{3}{10}$.

## SET 4 Extension

1 What is the 8th letter of the alphabet?

2 How many grams in $\frac{1}{2}$ kg?

3 Share $40 among 5 people.

4 How much is 3 kg of tomatoes at $3 per kg?

5 How much is $\frac{1}{2}$ kg meat at $9.00 per kg?

6 What is the value of 8 in 3783?

7 How many legs on 35 people?

8 What is 13 more than 267?

9 Does a pentagon have 6 sides?

10 4 m of material costs $12. How much for 7 m?

Mathematical Reasoning

11 Jessie threw 3 darts and scored 100. Name 3 different ways of scoring 100.

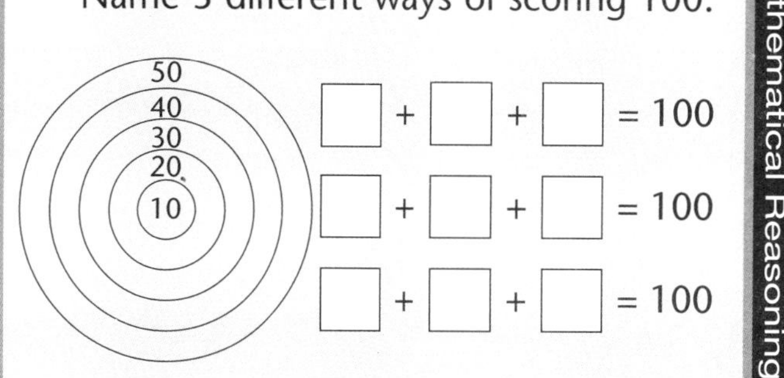


# Statistics and Probability Column graphs

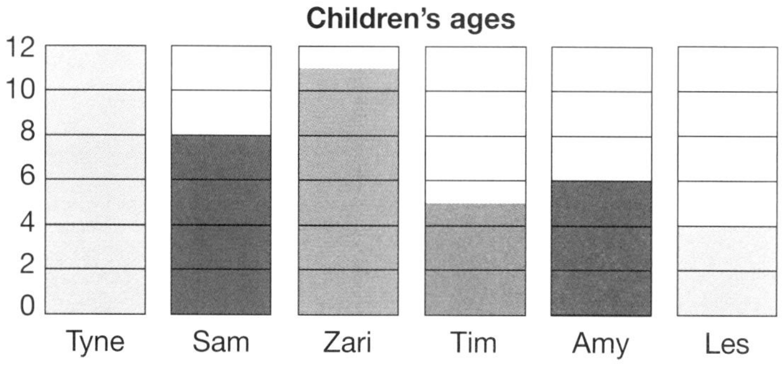


Answer the questions on the 6 children's ages.

1 How old is Les?

2 How old is Sam?

3 How old is Zari?

4 How much older is Tyne than Amy?

5 How much younger is Tim than Zari?

UNIT 5

# Number and Algebra

## SET 1 Basic

1 3 + ☐ = 10

2 8 + ☐ = 11

3 11 – 3

4 12 – ☐ = 8

5 2 × 10

6 3 × 2

7 4 × 6

8 ☐ × 5 = 20

9 What is the 2nd month of the year?

10 What is the product of 6 and 3?

11 What is the sum of 18 and 4?

12 What is the difference between 26 and 4?

13 How many minutes in 2 hours?

14 4372 = ☐ thousands + ☐ hundreds + ☐ tens + ☐ ones

15

## SET 2 Revising 3-digit addition

1

| | HUND | TENS | ONES |
|---|---|---|---|
| | 3 | 7 | 6 |
| + | 2 | 6 | 6 |
| | | | |

2

| | HUND | TENS | ONES |
|---|---|---|---|
| | 1 | 6 | 8 |
| + | 3 | 6 | 6 |
| | | | |

3

| | HUND | TENS | ONES |
|---|---|---|---|
| | 4 | 5 | 6 |
| + | 4 | 7 | 8 |
| | | | |

4

| | HUND | TENS | ONES |
|---|---|---|---|
| | 6 | 8 | 6 |
| + | 1 | 8 | 4 |
| | | | |

Add these numbers by making to a hundred first. E.g. 85 + 35 becomes 85 + 15 + 20 = 120.

5 95 + 55 =

6 88 + 42 =

7 196 + 34 =

8 296 + 44 =

9 391 + 59 =

10 380 + 55 =

11 488 + 62 =

# Space Symmetry

Complete these shapes using their line of symmetry.

# Number and Algebra

## SET 3 Multiples of 10

Complete these multiplications.

1 $6 \times 2$
2 $6 \times 20$
3 $8 \times 2$
4 $8 \times 20$
5 $6 \times 3$
6 $6 \times 30$
7 $7 \times 3$
8 $7 \times 30$
9 $4 \times 4$
10 $4 \times 40$
11 $7 \times 4$
12 $7 \times 40$
13 $5 \times 5$
14 $5 \times 50$
15 $6 \times 6$
16 $6 \times 60$

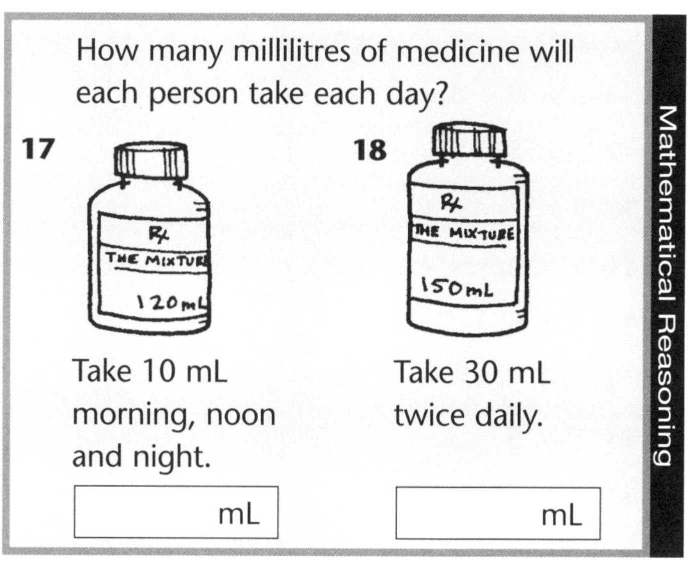


## SET 4 Extension

1 45 + 5 + 45
2 145 – 60
3 70 + 30
4 What is the product of 5 and 4?
5 What is the difference between 18 and 9?
6 3 lollies at 10c each
7 50 less than 1000
8 (24 – 4) + 2
9 Triple 4.
10 How many quarters in $1\frac{1}{4}$?
11 $\frac{1}{4}$ of 16
12 2340 + 300
13 Double 24.
14 4185 less one hundred
15 Write one thousand, three hundred and twenty-four in figures.
16 Complete the number web.

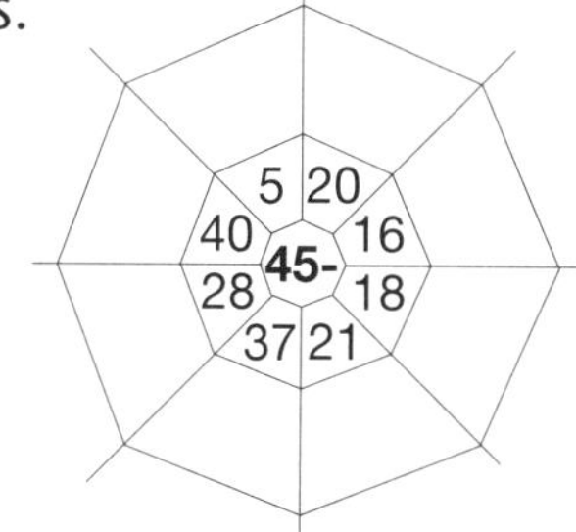


# Measurement Informal area units

Count the squares to work out the area of the following shaded shapes.

1
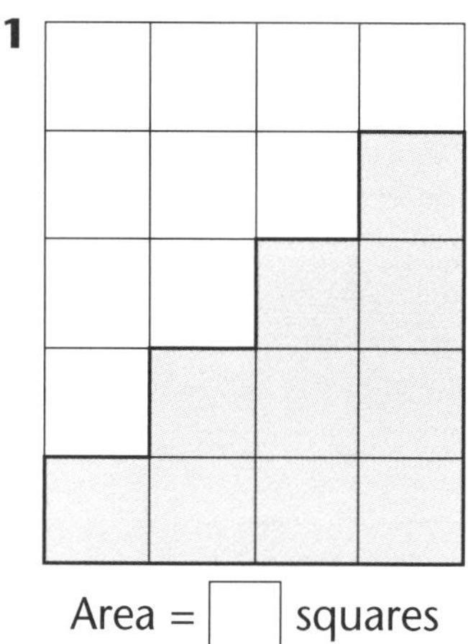
Area = ☐ squares

2
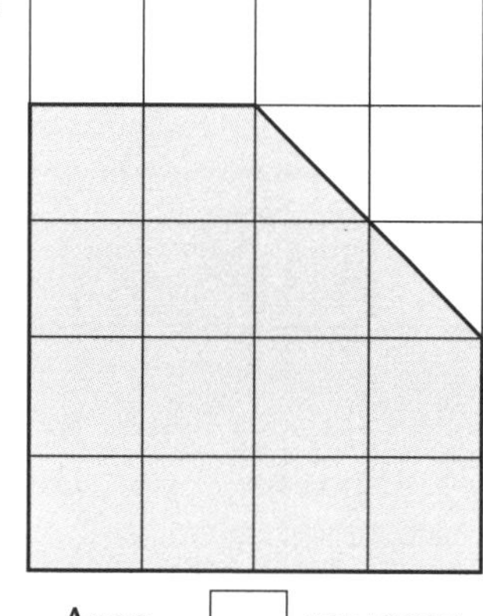
Area = ☐ squares

3
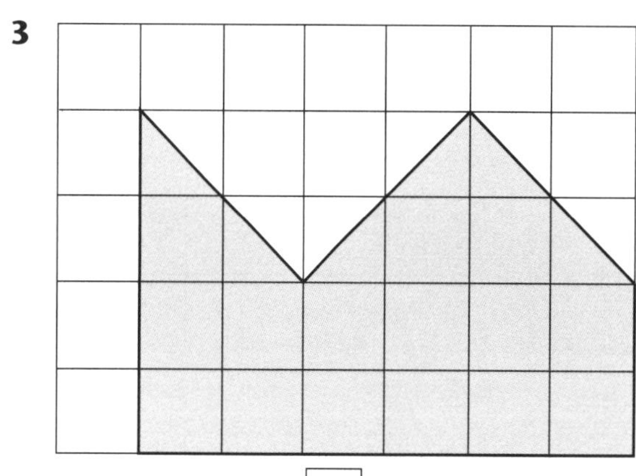
Area = ☐ squares

UNIT 6

# Number and Algebra

## SET 1 Basic

1 26 + 3
2 74 + 6
3 3 + 7 + 5
4 26 – 5
5 28 – 4
6 2 × 6
7 3 × 5
8 4 × 4
9 5 × 3
10 What is the product of 6 and 3?
11 19 take away 6
12 Triple 3.
13 Half of 48
14 Hours between 2 pm and 10 pm? _______
15 7 thousands + 6 hundreds + 3 tens + 7 ones

_______________

16

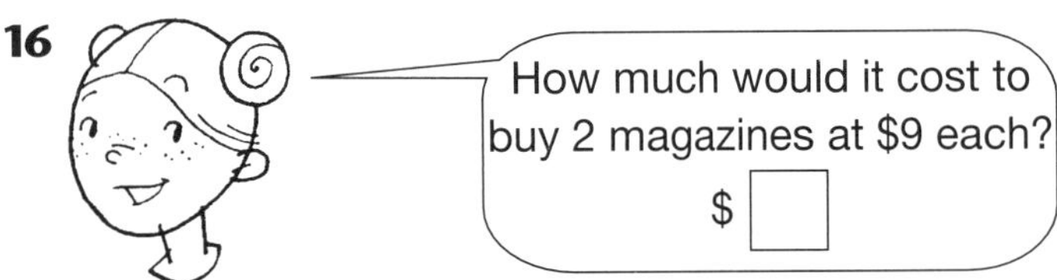


## SET 2 Multiplication facts

1

| | ×4 |
|---|---|
| 2 | |
| 4 | |
| 5 | |
| 6 | |
| 7 | |
| 8 | |

2

| | ×5 |
|---|---|
| 2 | |
| 3 | |
| 4 | |
| 6 | |
| 8 | |
| 9 | |

3

| | ×6 |
|---|---|
| 2 | |
| 4 | |
| 5 | |
| 6 | |
| 9 | |
| 8 | |

4 6 cakes at $4 each
5 7 books at $3 each
6 8 pens at $6 each
7 8 cakes at $5 each
8 3 × 4 + 2
9 5 × 5 + 5
10 9 × 2
11 9 × 20
12 4 × 3
13 4 × 30
14 7 × 5
15 7 × 50

# Space Angles

Shade all right angles yellow.

Shade all the angles smaller than a right angle blue.

Shade all angles larger than a right angle red.

# Number and Algebra

## SET 3 Jump strategy

To add 126 + 34

Think

126 + 30 + 4

1 38 + 26 =

2 65 + 34 =

3 126 + 62 =

4 177 + 21 =

5 161 + 37 =

6 232 + 46 =

7 314 + 65 =

8 413 + 78 =

9 527 + 55 =

10 638 + 64 =

11 347 + 48 =

12 466 + 28 =

13 Jack bought a TV for $518 and a stand for $69. How much did he spend altogether?

## SET 4 Extension

1 Share 70c between 2 people.

2 How many 10c coins make $4.00?

3 How much is 4 kg of onions at 55c per kg?

4 What is the value of 9 in 5945?

5 Estimate an answer to 1596 + 201.

6 Round 1594 to the nearest 10.

7 Round 1594 to the nearest 100.

8 Round 2680 to the nearest 1000.

9 How many hours are there from 6 am to 6 pm?

10 Multiply 6 by 4 and add 5.

11 9, ☐, 27, ☐, ☐, 54, ☐

12 What is the change from $3 if I spent $1.65?

13 Make the smallest number possible using 9, 6, 9, 6.

14 How much money did the children pay for their excursion? On Monday their teacher collected $40, on Tuesday she collected twice as much and on Wednesday she collected $120.

Mathematical Reasoning

# Measurement Mass in kilograms

Write the item that is heavier.

1 A large bucket full of water **or** a 1 kg mass

2 A TV set **or** a 1 kg mass

3 A pair of sunglasses **or** a 1 kg mass

4 A 2 kg bag of potatoes **or** a 1 kg mass

5 This book **or** a 1 kg mass

6 A 1 kg bag of feathers **or** a $\frac{1}{2}$ kg bag of rocks

7 995 g **or** 1 kg

UNIT 7

# Number and Algebra

## SET 1 Basic

1 5 + 6 + 8

2 9 − 7

3 4 × 2

4 3 × 2 + 4

5 What is the product of 7 and 2?

6 What is the sum of 17 and 7?

7 4 + ☐ = 11

8 Double 9.

9 ☐ × 4 = 20

10 What is the product of 7 and 9?

11 3 + 3 + 3

12 Triple 5.

13 27 + 8

14 What is half of 36?

15

## SET 2 Adding or multiplying in any order

Complete the sets of additions.

| | |
|---|---|
| 1 | 8 + 7 + 6 =<br>6 + 8 + 7 = |
| 2 | 20 + 80 + 70 =<br>80 + 70 + 20 = |
| 3 | 70 + 90 + 60 =<br>60 + 70 + 90 = |

Complete the sets of multiplications.

| | |
|---|---|
| 4 | 3 × 2 × 5 =<br>5 × 3 × 2 = |
| 5 | 2 × 4 × 5 =<br>4 × 5 × 2 = |
| 6 | 10 × 1 × 3 =<br>1 × 3 × 10 = |

Look for numbers that go together to make each addition easier.

7 75 + 69 + 25 =

8 87 + 29 + 13 =

9 16 + 17 + 14 + 13 =

10 85 + 65 + 15 + 35 =

# Statistics and Probability Picture graphs

**Transport to school**

| | |
|---|---|
| Bus | ☺☺☺☺☺☺☺ |
| Train | ☺☺☺☺ |
| Walk | ☺☺☺☺☺☺☺☺ |
| Car | ☺☺☺☺☺ |
| Bike | ☺☺☺ |

**Number of people**

| KEY | ☺ = 10 children |
|---|---|

1 Which is the most common form of transport?

_______________

2 Which is the least common form of transport?

_______________

3 How many more walkers are there than bike riders?

_______________

4 Which two groups have a combined total equal to the bus group?

_______________

# Number and Algebra

## SET 3 Thirds and sixths

Shade the given fraction.

**1** $\frac{1}{3}$

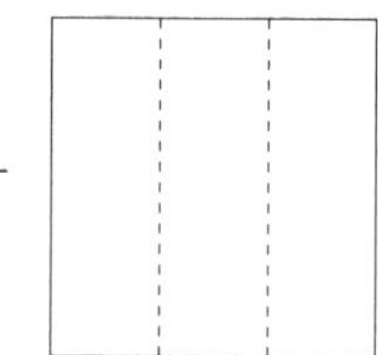

**2** $\frac{2}{3}$

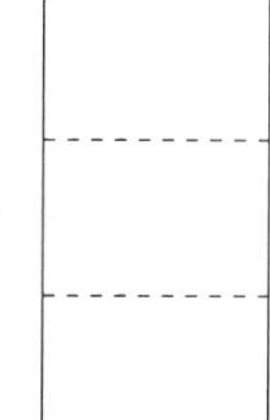

Shade the fraction of each group.

**3** $\frac{1}{3}$

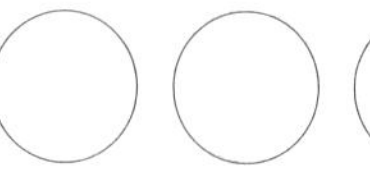

**4** $\frac{2}{3}$

Write true or false.

> greater than
< less than

**5** $\frac{1}{3} < \frac{1}{10}$ ________

**6** $\frac{1}{3} < \frac{1}{2}$ ________

**7** $\frac{1}{3} < \frac{1}{8}$ ________

**8** $\frac{1}{3} > \frac{1}{4}$ ________

## SET 4 Extension

**1** What is the total of 20, 60 and 40?

**2** Subtract 19 from 350.

**3** What is the value of 5 in 4517?

**4** Round 2790 to the nearest 100.

**5** 36 – 8 + 2

**6** 15 more than 149

**7** Share $35 among 5 people.

**8** Multiply 4 by 2 and add 6.

**9** $3.15 = ☐ cents

**10** 5 × 2 + 6

**11** 5 × (2 + 6)

**12** How many tens in 539?

**13**

Hawaiian

$1 per slice

Supreme

$1.20 per slice

Special

$1.10 per slice

Name 2 orders that cost $3.30.

________________

________________

Mathematical Reasoning

# Measurement Calculating area

Draw 2 different shapes that cover an area of 15 squares.

# Number and Algebra

## SET 1 Basic

**1** 4 + 3 + 6

**2** 7 + 2 + 1

**3** 19 – 3

**4** 21 km – 5 km

**5** 6 × 3

**6** 7 × 4

**7** 5 × 4

**8** 9 × 0

**9** The 5th month of the year is ☐.

**10** Sum of 15 and 4

**11** Share $12 among 6 people.

**12** Multiply 8 by 3.

**13** Halve 20c.

**14** What is the time 3 minutes after 4:15?

**15** 6 hundreds + 4 tens + 3 ones

**16**

## SET 2 Revising 3-digit subtraction

**1**

| | HUND | TENS | ONES |
|---|---|---|---|
| | 7 | 3 | 2 |
| – | 3 | 0 | 6 |
| | | | |

**2**

| | HUND | TENS | ONES |
|---|---|---|---|
| | 8 | 5 | 4 |
| – | 3 | 6 | 5 |
| | | | |

**3**

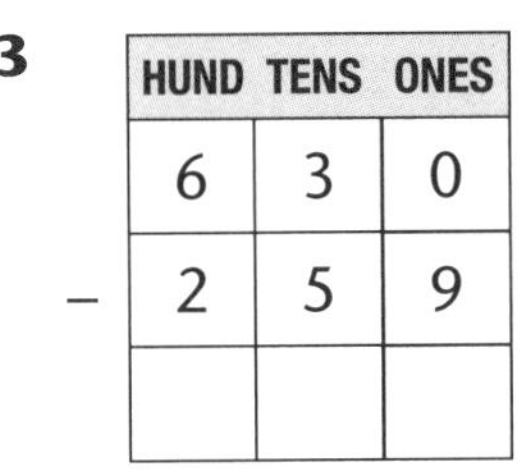

| | HUND | TENS | ONES |
|---|---|---|---|
| | 6 | 3 | 0 |
| – | 2 | 5 | 9 |
| | | | |

**4**

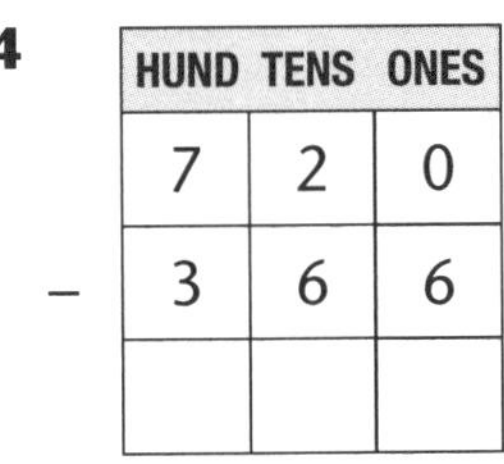

| | HUND | TENS | ONES |
|---|---|---|---|
| | 7 | 2 | 0 |
| – | 3 | 6 | 6 |
| | | | |

**5** 800 – 500

**6** 1800 – 700

**7** 480 – 60

**8** 857 – 30

**9** 946 – 25

**10** 7746 – 700

**11** Jim bought a new ring for $496. If the jeweller gave him a $55 discount, how much did he spend? 

# Number and Algebra Rounding to 10

Round off each number to the nearest 10.

| | Number | Nearest 10 |
|---|---|---|
| **1** | 18 | |
| **2** | 25 | |
| **3** | 32 | |
| **4** | 50 | |
| **5** | 74 | |
| **6** | 89 | |

Round off each addition to the nearest 10 to make an estimate.

| | Adding | Estimate |
|---|---|---|
| **7** | 22 + 19 = | 20 + 20 = 40 |
| **8** | 40 + 33 = | |
| **9** | 51 + 45 = | |
| **10** | 86 + 62 = | |
| **11** | 102 + 77 = | |

# Number and Algebra

## SET 3 Number patterns

Complete the number patterns.

1

| 5 | 10 | 15 | | | | | |
|---|---|---|---|---|---|---|---|

2

| 0 | 3 | 6 | | | | | |
|---|---|---|---|---|---|---|---|

3

| 20 | 30 | 40 | | | | | |
|---|---|---|---|---|---|---|---|

4

| 23 | 26 | 29 | | | | | |
|---|---|---|---|---|---|---|---|

5

| 216 | 218 | 220 | | | | | |
|---|---|---|---|---|---|---|---|

6

| 325 | 320 | 315 | | | | | |
|---|---|---|---|---|---|---|---|

7

| 500 | 450 | 400 | | | | | |
|---|---|---|---|---|---|---|---|

8

| 550 | 600 | 650 | | | | | |
|---|---|---|---|---|---|---|---|

Write the next number in the sequence.

**9** 245, 445, 645, ________

**10** 1231, 2231, 3231, ________

**11** 3107, 3307, 3507, ________

## SET 4 Extension

**1** 1 m and 27 cm = [ ] cm

**2** What is the 6th letter of the alphabet?

**3** How many $\frac{1}{4}$s in $1\frac{1}{2}$?

**4** Share 36 among 4.

**5** 2 kg = [ ] g

**6** Value of 2 in 3286

**7** Multiply 3 by 2 and add 56.

**8** 3 thousands + 2 hundreds + 6 ones

**9** 2427 + 70

**10** Round 1297 to the nearest 100.

**11** 27 more than 1351

**12** If 3 kg cost $9, how much would 4 kg cost?

**13** How many hours from noon to midnight?

**14** How many 10c coins make $2.80?

**15** Write 1400 in words. ________

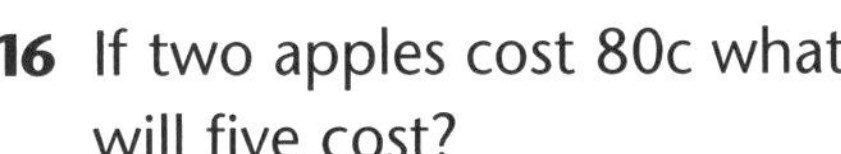

**16** If two apples cost 80c what will five cost?

**17** Round off to estimate an answer to 217 + 31.

# Measurement Litres

Match a container to each quantity.

**1** 1 litre [ ]

**2** 2 litres [ ]

**3** 4 litres [ ]

**4** 10 litres [ ]

**5** 20 litres [ ]

UNIT 9

# Number and Algebra

## SET 1 Basic

1 6 + 4 + 11

2 9 + 1 + 10

3 18 – 16

4 19 – 1

5 5 × 4

6 11 × 3

7 8 × 4

8 10 × 2

9 The 7th month of the year is ☐.

10 What is the product of 6 and 8?

11 Share 8 between 2.

12 Add 4 and 21.

13 Triple 1.

14 15 minutes after 1:10 pm

15 

## SET 2 Multiplication facts

1

| ×3 | |
|---|---|
| 2 | |
| 4 | |
| 5 | |
| 6 | |
| 7 | |
| 8 | |

2

| ×7 | |
|---|---|
| 2 | |
| 3 | |
| 4 | |
| 6 | |
| 8 | |
| 9 | |

3

| ×5 | |
|---|---|
| 2 | |
| 4 | |
| 5 | |
| 6 | |
| 9 | |
| 8 | |

4 6 books at $7 each

5 7 gluesticks at $3 each

6 8 presents at $6 each

7 8 cakes at $7 each

8 3 × 7 + 20

9 5 × 7 + 50

10 9 × 7 + 7

11 7 rows of 6 trees

12 9 rows of 5 plants

13 10 rows of 7 trees

14 3 × 7

15 3 × 70

# Space Location

**Treasure island**

Mark a line on the map to where the skull is by following the directions.

Mathematical Reasoning

# Number and Algebra

## SET 3 Expanding numbers

Expand these numbers.

1 4736

4000 + ☐ + ☐ + ☐

2 5674

☐ + ☐ + ☐ + ☐

Write the number before and after.

3 ________ 1234 ________

4 ________ 5678 ________

5 ________ 2853 ________

Order these numbers from smallest to largest.

| | | | | |
|---|---|---|---|---|
| 6 | 73 567 | 1299 | 44 974 | |
| 7 | 35 675 | 13 679 | 24 674 | |
| 8 | 44 376 | 57 643 | 66 743 | |

9 Add 10 to 8050.

10 Add 20 to 3568.

11 Add 30 to 2371.

12 Add 200 to 5870.

13 Add 300 to 1256.

## SET 4 Extension

1 How much do nine pens at $1.10 cost?

2 What is the product of 8 and 6?

3 What is the sum of 40, 130 and 100?

4 Complete this addition pattern:

1, 3, 6, ☐, 15

5 $10 \times 4 + 6$

6 2295 cents = $ ☐

7 3 tens plus 5 tens

8 How many tens in 1486?

9 What is the difference between 89 and 65?

10 87 more than 1303

11 $\frac{1}{4}$ of 64

12 Share $1.50 among 3 people.

13 If 6 cost $3, how much would 8 cost?

14 LXXII =

15 Write 1964 in words.________________________

________________________________

16 Each of 4 tables has 8 people sitting at them. How many people are there altogether?

# Statistics and Probability Chance

1 Where do you think you will swim next? Connect each situation to its likelihood.

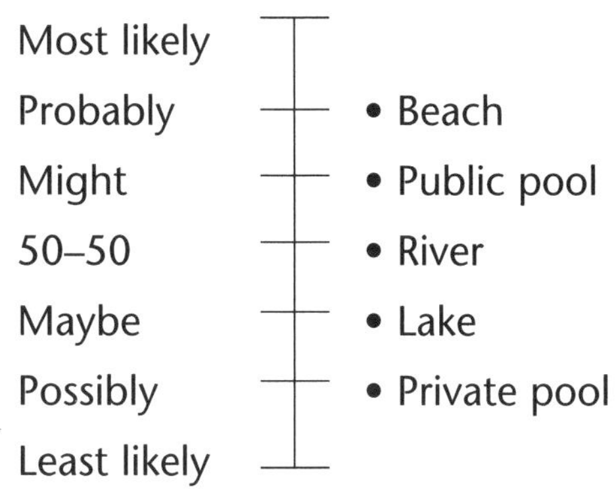


2 What is the likelihood of the hairdresser's shop being open now? Match each sign to a chance card.

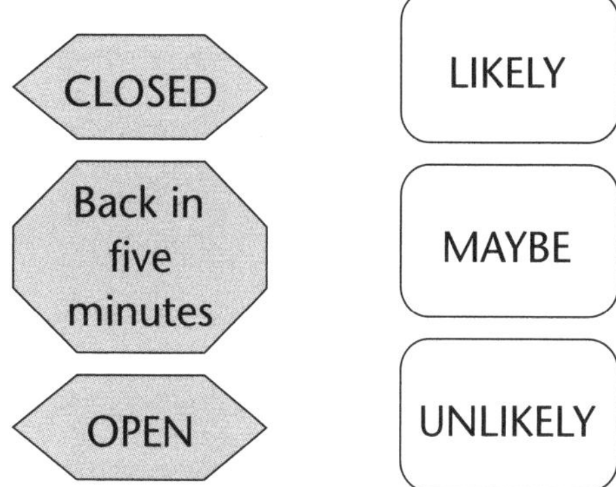

UNIT 10

# Number and Algebra

## SET 1 Basic

1 6 cm + 3 cm + 4 cm

2 8 − 7

3 36 − 4

4 7c + 3c + 10c

5 5 × 6

6 3 × 4

7 2 × 7

8 4 × 8

9 What is the 10th month?

10 What is the difference between 26 and 4?

11 What is the sum of 18 and 10?

12 Triple 6.

13 Double 16.

14 How many hours between 3 pm and 12 midnight?

15 7 thousands + 1 hundred + 3 tens + 7 ones

16 

I am half my brother's age. How old am I if he is 18?

☐ years

## SET 2 Multiplication strategies

To multiply by 4, double then double again.

Complete these multiplications.

1 7 × 4 =

2 9 × 4 =

3 11 × 4 =

4 21 × 4 =

5 40 × 4 =

6 30 × 4 =

To multiply by 5, multiply by 10 then halve.

7 14 × 5 =

8 22 × 5 =

9 26 × 5 =

10 18 × 5 =

11 46 × 5 =

12 24 × 5 =

Use your own strategies to solve these multiplications.

13 12 × 6 =

14 11 × 5 =

15 32 × 6 =

16 12 × 8 =

# Space Tessellations

Put a tick on all shapes that tessellate and put a cross on those that do not.

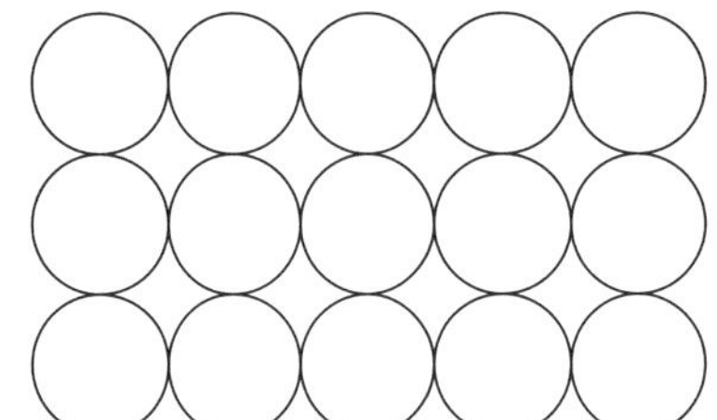

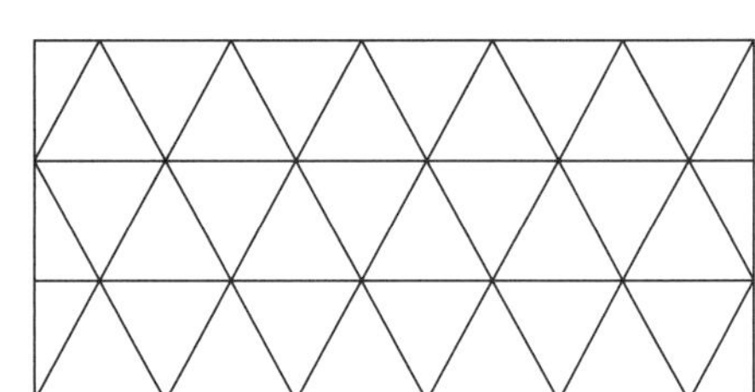

# Number and Algebra

## SET 3 Equivalent fractions

Shade then write an equivalent fraction for each one given.

**1** 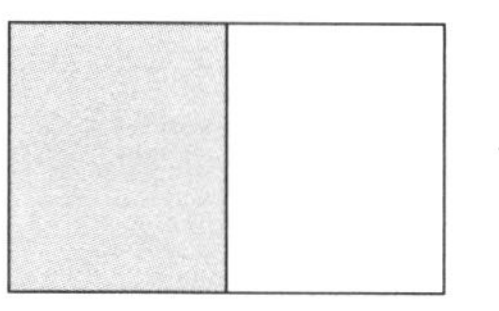 $\frac{1}{2}$

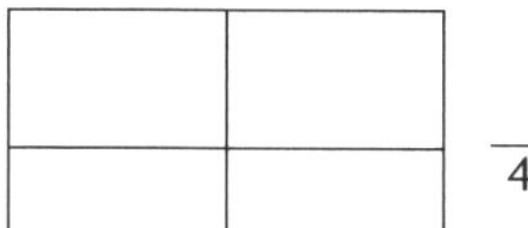 $\frac{\quad}{4}$

**2** 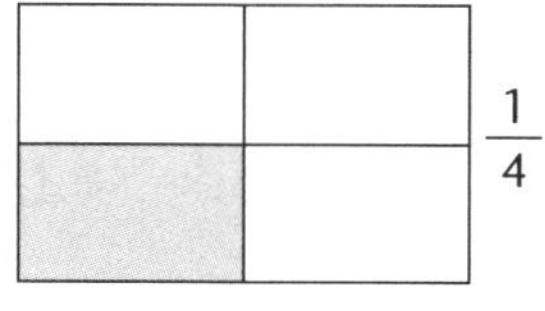 $\frac{1}{4}$

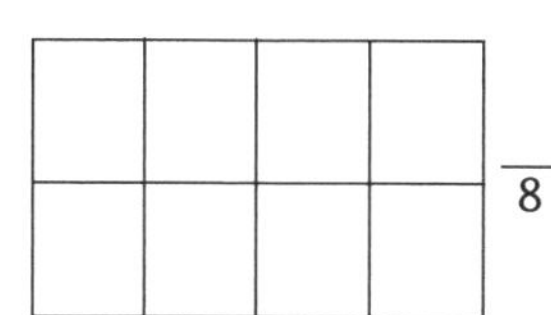 $\frac{\quad}{8}$

**3** 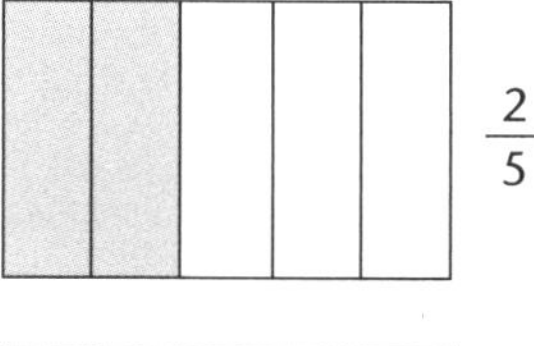 $\frac{2}{5}$

$\frac{\quad}{10}$

**4** 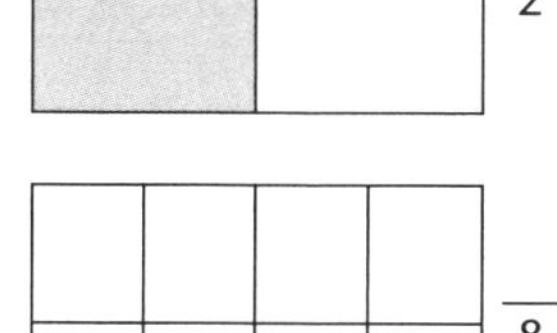 $\frac{1}{2}$

$\frac{\quad}{8}$

Write *true* or *false*.

**5** $\frac{1}{2} = \frac{5}{10}$ __________

**6** $\frac{1}{2} = \frac{5}{8}$ __________

**7** $\frac{1}{4} > \frac{1}{2}$ __________

**8** $\frac{1}{5} > \frac{1}{10}$ __________

**9** $\frac{1}{10} < \frac{1}{2}$ __________

**10** $\frac{7}{10} < \frac{1}{10}$ __________

## SET 4 Extension

**1** $5 \times 11 + 6$

**2** What is the cost of 12 pens at $7 each?

**3** Estimate an answer to $903 + 38$.

**4** 9 hundreds + 6 tens + 5 ones

**5** $6 \times 12 + 3$

**6** What is the sum of 40, 30 and 250?

**7** $\frac{1}{4}$ of 100

**8** What is the difference between 95 and 16?

**9** Write the largest number you can using 5, 2, 8, 7.

**10** $6 \times 8 - 4$

**11** How many tens in 4270?

**12** Use the clues to complete the graph.

| | 0 4 8 12 16 20 24 |
|---|---|
| Girls | (bar to 24) |
| Boys | |
| Women | |
| Men | |

Boys: One half the number of girls.
Women: One third the number of girls.
Men: One half the number of women.

# Measurement Millilitres

**1** How many millilitres in 1 L?

**2** How many millilitres in $\frac{1}{2}$ L?

**3** How many millilitres in 4 L?

**4** How many millilitres in $\frac{1}{4}$ L?

**5** How many millilitres in $3\frac{1}{2}$ L?

**6** How many millilitres equals 5 L?

**7** How much is left in a half-litre bottle if 250 mL is poured out?

**8** I have half a litre of milk. How many more millilitres do I need to make 1 L?

# Number and Algebra

## SET 1 Basic

1 6 + 2 + 8

2 20 + 30 + 8

3 19 − 11

4 25 − 14

5 6 × 2

6 11 × 0

7 8 × 5

8 7 × 6

9 How many months in half a year?

10 What is the product of 5 and 9?

11 What is the sum of 5 and 9?

12 What is the difference between 5 and 9?

13 Triple 5.

14 What time is it?

15

What is the total of 9 hundreds + 6 tens + 3 ones?

## SET 2 Division strategies

Use halving skills to solve the divisions.

1 40 ÷ 2

2 18 ÷ 2

3 Divide 36 by 2.

4 Divide 64 by 2.

Use the halve and halve again strategy to solve these divisions.

5 24 ÷ 4

6 40 ÷ 4

7 Divide 32 by 4.

8 Share 20 lollies among 4 people.

9 Share $60 among 4 people.

10 Dennis had 50 football cards but shared them between himself and four friends. How many did each child receive?

11 48 cows were shared between 4 paddocks. How many cows were put in each paddock?

# Statistics and Probability Picture graphs

**Pocket money per week**

Gina     

Max   

Zoe  

Lee 

Alex      

**Number of dollars**

1 How much does Gina earn per week? ________

2 Who earns the most money per week? ________

3 How much does Lee earn per week? ________

4 How much do Zoe and Max earn between them? ________

5 How much more does Alex earn compared to Max? ________

6 What is the total amount of money earned by the children? ________

# Number and Algebra

## SET 3 Multiplication facts (8s)

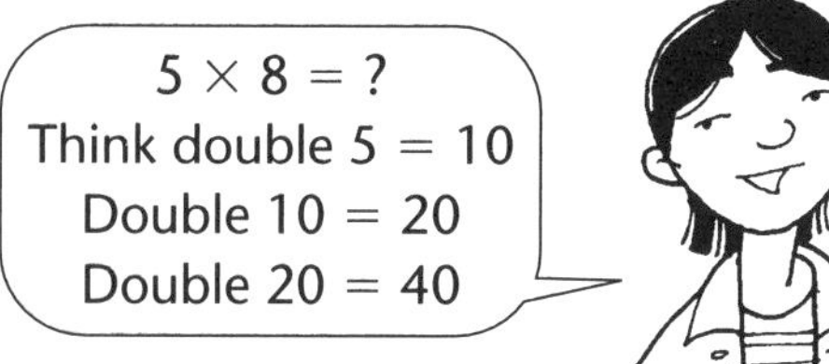


Use the double, double, double again strategy to solve these questions.

| | | | |
|---|---|---|---|
| **1** | $3 \times 8 =$ ☐ | **2** | $5 \times 8 =$ ☐ |
| **3** | $7 \times 8 =$ ☐ | **4** | $9 \times 8 =$ ☐ |
| **5** | $10 \times 8 =$ ☐ | **6** | $6 \times 8 =$ ☐ |
| **7** | $4 \times 8 =$ ☐ | **8** | $11 \times 8 =$ ☐ |
| **9** | $2 \times 8 =$ ☐ | **10** | $12 \times 8 =$ ☐ |

Complete the multiplication targets.

**11**

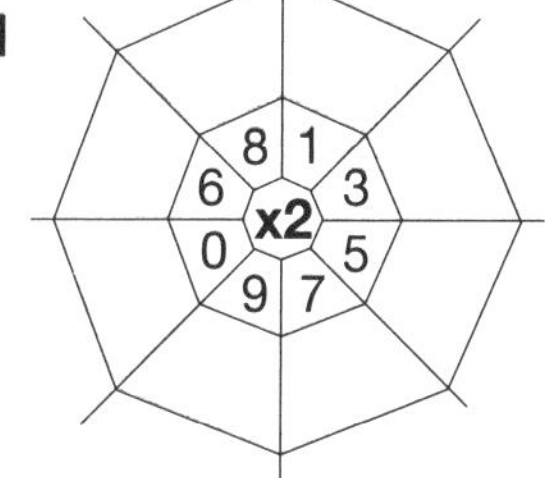


**12**

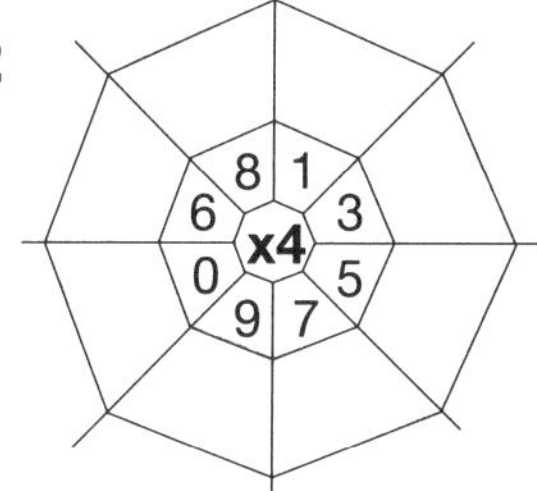


**13**

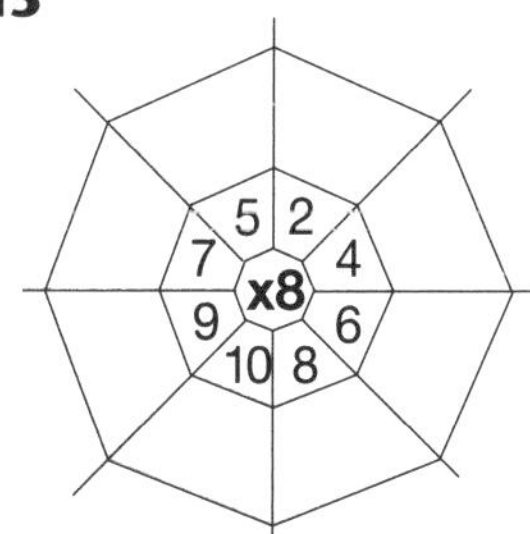


**14**

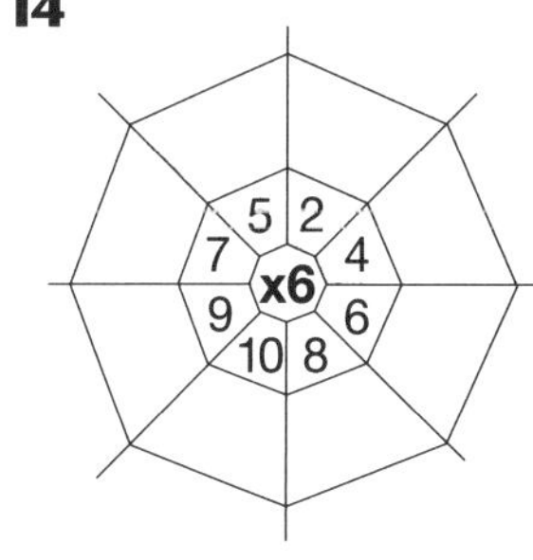


## SET 4 Extension

**1** 350, 356, 362, ☐

**2** 800 + 156

**3** Write the largest number you can using 1, 3, 7, 6.

**4** Share $100 among 4 people.

**5** $\frac{1}{2}$ of 56

**6** Round 7201 to the nearest 1000.

**7** What is the value of 3 in 3420?

**8** Write seven thousand and twenty-six in numerals.

**9** How many tens in 2436?

**10** How many faces does a cone have?

**11** How many minutes are there from 3 pm to 7:30 pm?

**12** How much is 8 kg of vegetables at $7 per kilogram?

**13** How much is $4\frac{1}{2}$ kg of meat at $8 per kilogram?

Mathematical Reasoning

**14** 8 people were supposed to share the $80 hire fee for a tennis court but Prani forgot her money. How much did Carly pay if she paid her share as well as Prani's share?

# Measurement Time units

**1** How many seconds in one minute? ☐

**2** How many minutes in one hour? ☐

**3** How many hours in one day? ☐

**4** How many days in a week? ☐

**5** How many days in a fortnight? ☐

**6** How many days in a year? ☐

Select the best unit of time to record these activities (seconds, minutes, hours, days, weeks).

**7** A netball game ☐

**8** A marathon run ☐

**9** Boiling an egg ☐

**10** Putting a stamp on an envelope ☐

# Number and Algebra

## SET 1 Basic

1 $8 + 4 + 3$

2 $8 + 8 + 8$

3 $24 - 10$

4 $35 - 6$

5 $7 + 7 + 7$

6 $4 + 4 + 4$

7 $10 \times 7$

8 $4 \times 9$c

9 14 kg + 5 kg

10 $18 + 4$

11 $5 + \square = 20$

12 $31 - 1$

13 $25 - 6$

14 $16 - \square = 12$

15 6 hundreds + 3 tens + 9 ones

16

| Rulers 60c | Pens $2 | Pencils 40c |
|---|---|---|

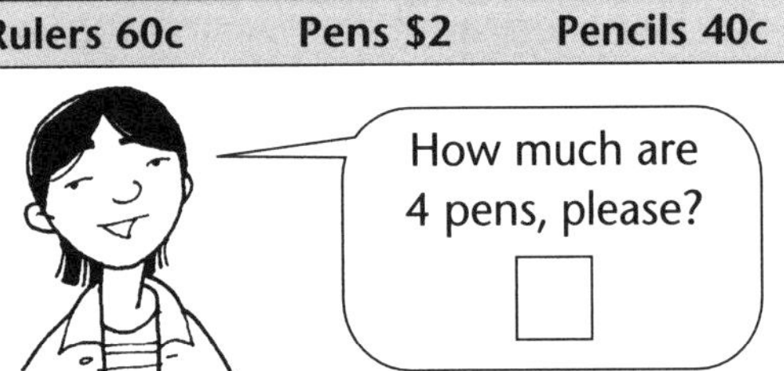


## SET 2 Multiplying mentally

Multiply the tens, then multiply the ones. $4 \times 35$ becomes $4 \times 30 + 4 \times 5 = 140$

1 $43 \times 5$

2 $27 \times 4$

3 $38 \times 6$

4 $25 \times 5$

5 $31 \times 4$

6 $14 \times 8$

7 $36 \times 5$

8 $42 \times 4$

9 $37 \times 6$

10 $63 \times 2$

**Mathematical Reasoning**

Round the 2-digit numbers to the nearest ten to give an approximate answer.

11 $29 \times 4 =$

12 $42 \times 5 =$

13 $68 \times 7 =$

# Space Drawing three-dimensional objects

Each dot represents a corner of a pyramid. Draw a line between each dot and the apex dot, then draw lines between the dots for the base. Name the shape of each pyramid base.

1

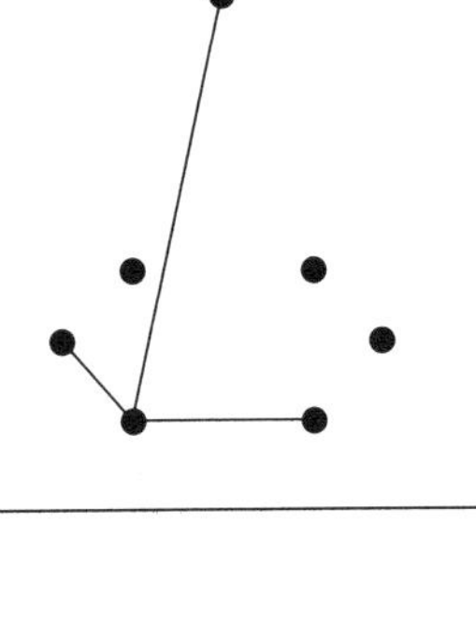

2

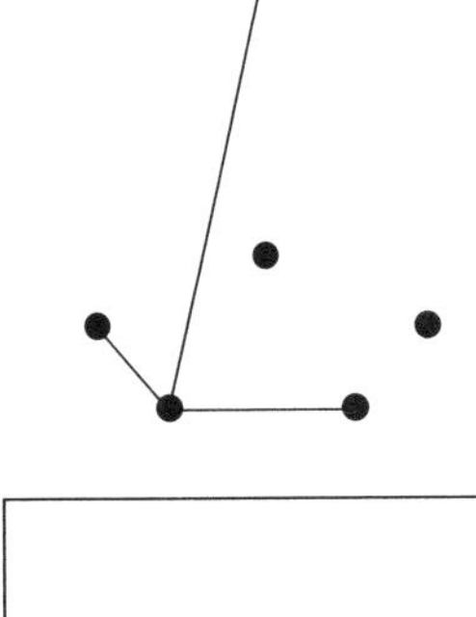

3

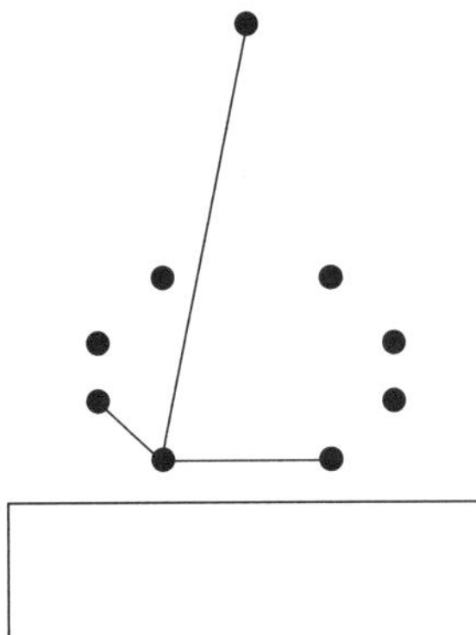

# Number and Algebra

## SET 3 Mixed numerals

Colour the shapes to match the mixed numerals.

1 $1\frac{2}{5}$

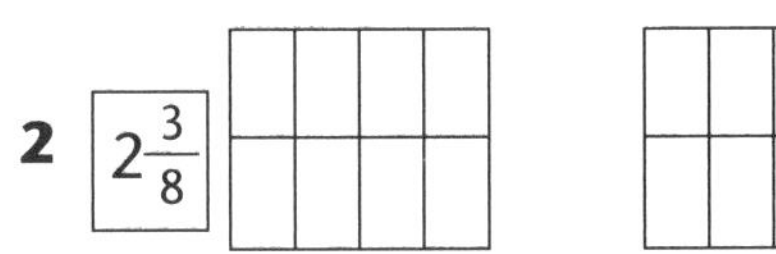

2 $2\frac{3}{8}$

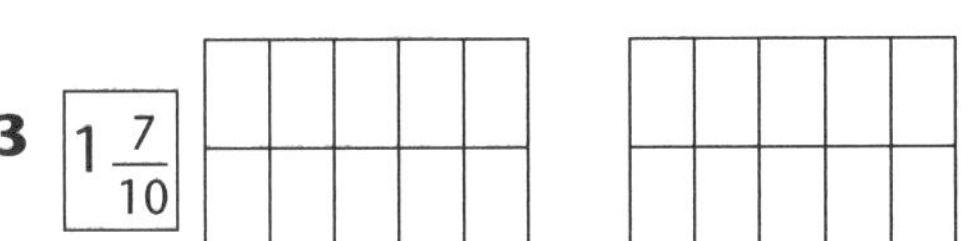

3 $1\frac{7}{10}$

4 $3\frac{1}{4}$

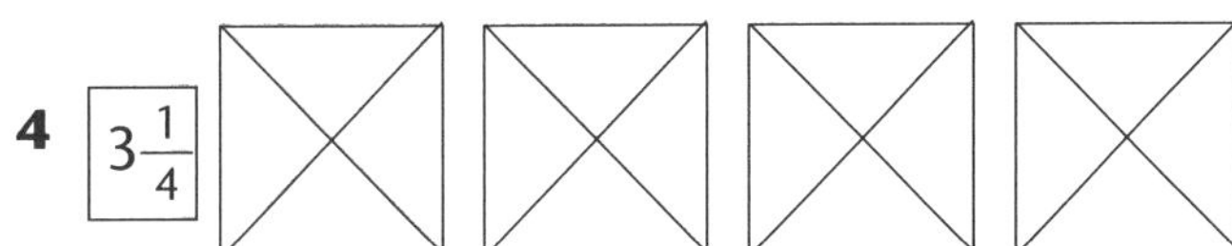

Write the next mixed numeral in the sequence.

5 1 $1\frac{1}{4}$ $1\frac{2}{4}$ ☐

6 $3\frac{1}{5}$ $3\frac{2}{5}$ $3\frac{3}{5}$ ☐

7 2 $2\frac{1}{2}$ 3 ☐

8 $1\frac{5}{10}$ $1\frac{6}{10}$ $1\frac{7}{10}$ ☐

## SET 4 Extension

1 $2 \times 4 + 10$

2 39 more than 261

3 Halfway between 40 and 50

4 Estimate an answer to 1401 + 59.

5 \$1.64 + \$5.00

6 375, 390, 405, ________ , ________

7 Two positions before 35th.

8 How many legs on 3 chairs with 3 people on them?

9 Share \$85 among 5 people.

10 If 4 shirts cost \$80, how much would 10 cost?

11 What is the value of 4 in 6400?

12 2186 cents = \$ ☐

13 399 take away 2 tens

14 Subtract $2 \times 10$ from 600.

15 What is half of 84?

16 $18 \div 3 + 87$

17 Find the quotient of 132 and 11.

# Measurement Grams

Find the difference in mass between:

1 a can of baked beans and a can of apples. ☐

2 a jar of jam and a can of baked beans. ☐

3 a bag of sweets and a packet of tea. ☐

4 a jar of jam and a can of apples. ☐

5 a jar of jam and a bag of sweets. ☐

# Number and Algebra

## SET 1 Basic

**1** 12 − 3

**2** 16 − 4

**3** 8 + 20

**4** 23 + 7

**5** 4 × 8

**6** 6 × 8

**7** 3 × 8

**8** 7c + 7c + 7c

**9** Subtract 6 from 37.

**10** \$3.86 = ☐ c

**11** How many 5c coins in 35c?

**12** How many threes in 15?

**13** What is the sum of 16 and 9?

**14** 1376 = ☐ thousand + ☐ hundreds + ☐ tens + ☐ ones

**15**

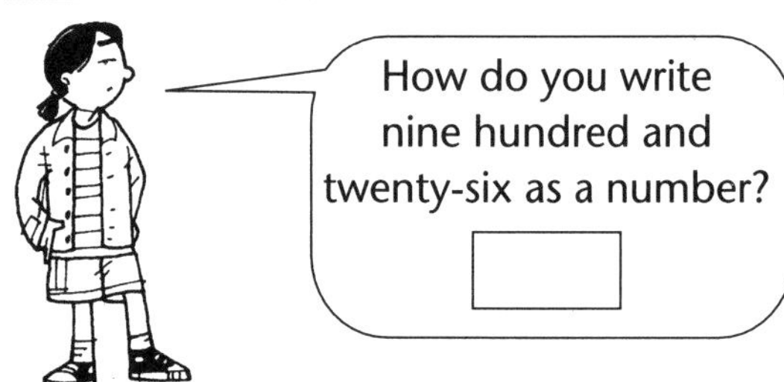


## SET 2 Compensation strategy

326 + 39 =
Think!

326 + 40 = 366
Then take away the 1 that
was used for rounding 39 to 40.
366 − 1 = **365**

**1** 243 + 38 =

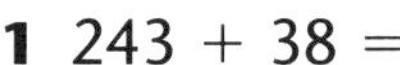

**2** 346 + 29 =

**3** 421 + 47 =

**4** 527 + 58 =

**5** 341 + 89 =

**6** 623 + 57 =

**7** 713 + 69 =

**8** Place the digits 6, 12, 18, 24 and 30 in the boxes so that the sum of the numbers is the same when added horizontally and vertically.

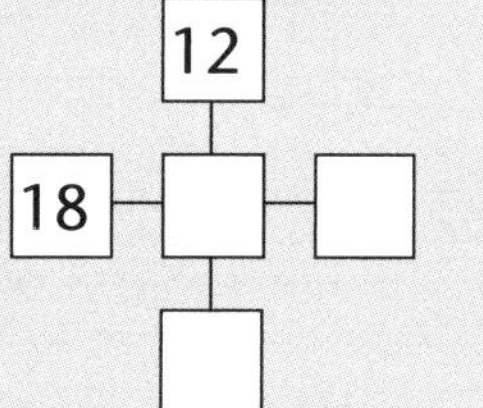


Mathematical Reasoning

# Measurement Metres

Would you measure the following in **cm** or **m**?

**1** The height of a house ______

**2** The length of a diary ______

**3** The length of a bus ______

**4** The length of an envelope ______

**5** The length of a swimming pool ______

Convert each metre measurement into centimetres.

**6** 2 m = ______cm

**7** 5 m = ______cm

**8** 9 m = ______cm

**9** $1\frac{1}{2}$ m = ______cm

**10** $1\frac{1}{4}$ m = ______cm

# Number and Algebra

## SET 3 Rounding to 100

Round off each number to the nearest 100.

| | Number | Nearest 100 |
|---|---|---|
| 1 | 976 | |
| 2 | 823 | |
| 3 | 1213 | |
| 4 | 1787 | |
| 5 | 2356 | |
| 6 | 7599 | |
| 7 | 4207 | |

Round off each addition to the nearest 100 to make an estimate.

| | Adding | Estimate |
|---|---|---|
| 8 | 296 + 703 = | 300 + 700 = 1000 |
| 9 | 237 + 759 = | |
| 10 | 425 + 899 = | |
| 11 | 901 + 989 = | |
| 12 | 876 + 797 = | |
| 13 | 1023 + 976 = | |

## SET 4 Extension

1 (6 + 2) × 4

2 Share 36 lollies among 7 people.

3 4325 + 500

4 6 rulers at 30c each

5 What is the value of 6 in 9652?

6 What is 19 less than 36?

7 Add $0.50 to $5.60.

8 How much is 9 kg of meat at $15 per kg?

9 $\frac{1}{4}$ of 24

10 How many halves in 6 wholes?

11 How much is $3\frac{1}{2}$ kg of sausages at $8 per kilogram?

12 What is the sum of 168 and 232?

13 Share $49 among 7 people.

14 402, 399, ☐, ☐, ☐, 387

**Mathematical Reasoning**

15 How old is Stephen if he is $\frac{1}{2}$ the age of Anne, who is $\frac{1}{4}$ the age of Bob? Bob is 40 years old.

# Statistics and Probability Most likely/least likely

Spinner A

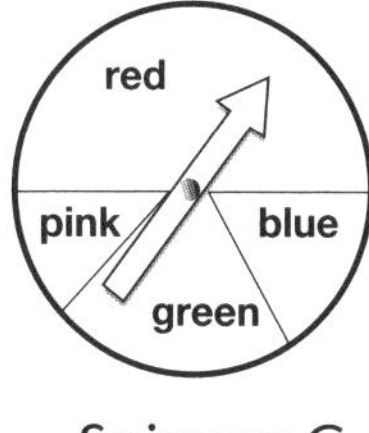


Spinner B

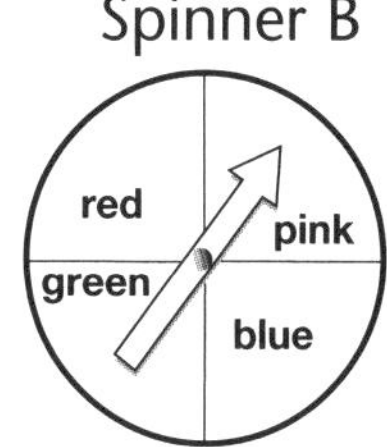


Spinner C

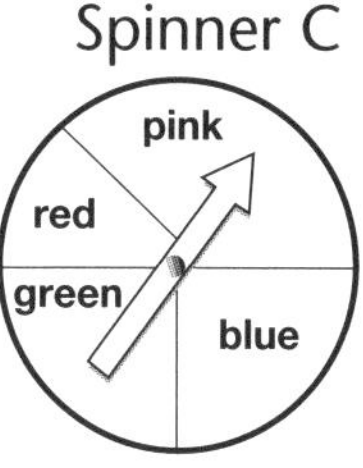


1 Which spinner is most likely to land on red? ☐

2 Which spinner is least likely to land on red? ☐

3 Which spinner is most likely to land on pink? ☐

4 Which spinner has equal chance of landing on any colour? ☐

5 Which spinner is least likely to land on pink? ☐

UNIT 14

# Number and Algebra

## SET 1 Basic

1 15 + 7

2 4 + 8 + 8

3 21 − 9

4 37 cm − 5 cm

5 7 × 3

6 5 × 3

7 4 × 4 + 5

8 Five lollies at 10c each

9 18 kg + 7 kg

10 Half of 30

11 How many days in 3 weeks?

12 Triple 8.

13 How many hours between 1 am and midday?

14 What is the difference between 33 and 5?

15 How much do 2 tickets to 'Ghost Dusters' cost?

| ADMIT ONE |
|---|
| GHOST DUSTERS $5.50 |

## SET 2 Addition

1

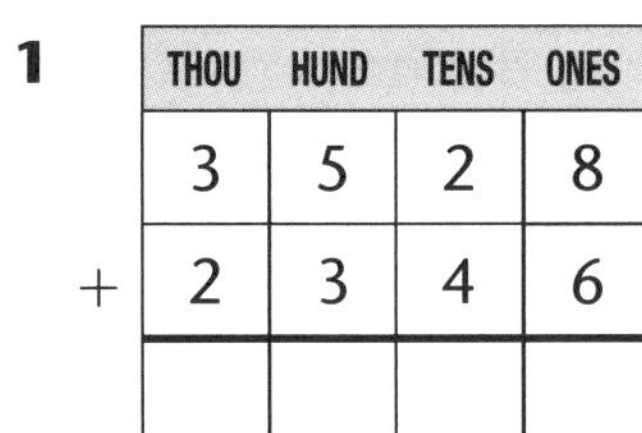

| | THOU | HUND | TENS | ONES |
|---|---|---|---|---|
| | 3 | 5 | 2 | 8 |
| + | 2 | 3 | 4 | 6 |
| | | | | |

2

| | THOU | HUND | TENS | ONES |
|---|---|---|---|---|
| | 1 | 3 | 5 | 7 |
| + | 7 | 3 | 4 | 7 |
| | | | | |

3

| | THOU | HUND | TENS | ONES |
|---|---|---|---|---|
| | 3 | 2 | 5 | 8 |
| + | 5 | 3 | 6 | 4 |
| | | | | |

4

| | THOU | HUND | TENS | ONES |
|---|---|---|---|---|
| | 3 | 6 | 8 | 7 |
| + | 3 | 2 | 4 | 5 |
| | | | | |

5

| | THOU | HUND | TENS | ONES |
|---|---|---|---|---|
| | 2 | 8 | 6 | 6 |
| + | 1 | 5 | 6 | 3 |
| | | | | |

6

| | THOU | HUND | TENS | ONES |
|---|---|---|---|---|
| | 3 | 7 | 8 | 2 |
| + | 5 | 7 | 6 | 3 |
| | | | | |

Supply the missing addends.

7

| | | | | |
|---|---|---|---|---|
| | 3 | 4 | 5 | ☐ |
| + | 2 | ☐ | 3 | 6 |
| | 6 | 2 | 8 | 9 |

8

| | | | | |
|---|---|---|---|---|
| | 2 | ☐ | 4 | 6 |
| + | 3 | 5 | ☐ | ☐ |
| | 6 | 3 | 1 | 8 |

# Number and Algebra Hundredths

Shade the hundredths grids to represent the hundredths.

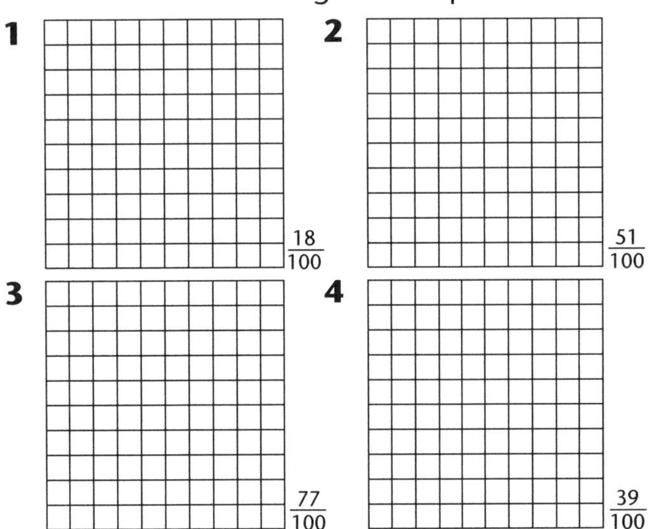

1 $\frac{18}{100}$

2 $\frac{51}{100}$

3 $\frac{77}{100}$

4 $\frac{39}{100}$

Trent measured 4 items using a metre ruler then recorded their measurements on a table.

Record each measurement as a fraction out of 100 cm (1 m).

| | | |
|---|---|---|
| 5 | 29 centimetres | $\frac{\quad}{100}$ |
| 6 | 45 centimetres | — |
| 7 | 80 centimetres | — |
| 8 | 92 centimetres | — |

# Number and Algebra

## SET 3 Division

1 Divide 24 by 4.

2 Divide 24 by 6.

3 Share 24 among 8.

4 Divide 32 by 8.

5 Divide 24 by 12.

6 Divide 36 by 9.

7 Share 100 among 10.

8 Share 81 among 9.

9 Share 36 among 6.

10 $18 \div 6$

11 $27 \div 3$

12 28 lollies are shared between 7 children.
They receive ☐ lollies each.

13 24 marbles are shared between 8 children.
They receive ☐ marbles each.

14 49 shapes are shared between 7 groups.
Each group receives ☐ shapes.

15 81 stickers are shared between 9 children.
Each child receives ☐ stickers.

## SET 4 Extension

1 Write the largest number you can using 2, 0, 9, 6.

2 How many sides on 5 pentagons?

3 How many minutes in $3\frac{1}{2}$ hours?

4 Share $28 equally between 4 people.

5 331, 340, 349, ☐, ☐, 376

6 If 6 hats cost $30, how much do 8 cost?

7 Add $1.50 to $4.25.

8 21 places behind 415th

9 How many halves in 8 wholes?

10 List the factors of 24.

11 How much does 3 kg of meat cost at $16 per kg?

12 How many legs on 4 bees and 7 dogs?

13 What is half of 660?

**Mathematical Reasoning**

14 Jennifer is trying to solve $8 \times 6$ on her calculator but the 6 button is missing.
Place numbers in the boxes so that her problem is solved.

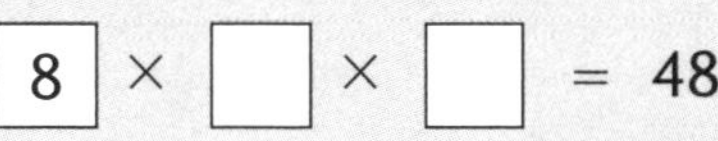


# Measurement The square centimetre

1 Work out the area of this shape. ☐ $cm^2$

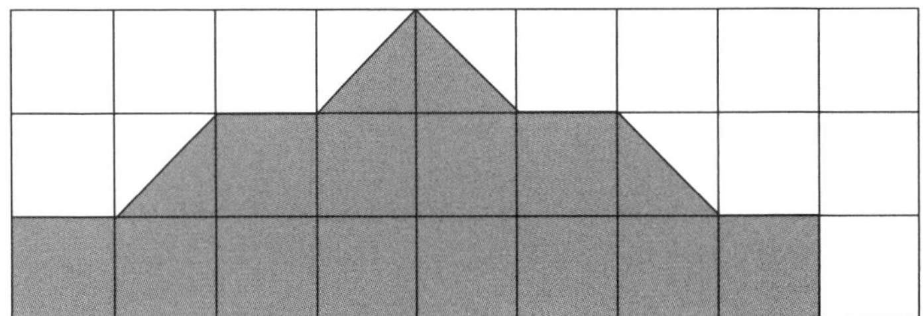

2 Draw a shape with an area of 24 $cm^2$.

UNIT
15

# Number and Algebra

## SET 1 Basic

1 $18 + 4$

2 $6 \times 7$

3 $5 \times 3 - 1$

4 $\$6 \times 4$

5 $3^2$

6 $4 + 4 + 4 + 4$

7 How much are 8 lollies at 5c each?

8 91 less 3

9 64 kg − 9 kg

10 What is the product of 11 and 4?

11 What is the difference between 21 and 3?

12 Half of $4 \times 5$

13 Subtract 6 from 21.

14 3:30 pm + 25 mins

15 ☐ hundreds + ☐ tens + ☐ ones = 695

16

How much change would I receive from $1 if I spent 85c? ☐

## SET 2 Multiplication facts (9s)

To multiply by 9 you can multiply by 10 then take away the number you multiplied.
$8 \times 9 =$
$8 \times 10 = 80 - 8 = 72$

Answer the multiplications.

| | | | |
|---|---|---|---|
| 1 | $2 \times 9 =$ ☐ | 6 | $4 \times 9 =$ ☐ |
| 2 | $3 \times 9 =$ ☐ | 7 | $6 \times 9 =$ ☐ |
| 3 | $1 \times 9 =$ ☐ | 8 | $8 \times 9 =$ ☐ |
| 4 | $5 \times 9 =$ ☐ | 9 | $9 \times 9 =$ ☐ |
| 5 | $10 \times 9 =$ ☐ | 10 | $7 \times 9 =$ ☐ |

Complete the multiplication targets.

11 (x7: 8, 1, 3, 5, 7, 9, 0, 6)

12 (x5: 8, 1, 3, 5, 7, 9, 0, 6)

# Space Identifying shapes

1 Put a cross on the rhombus.

2 Draw stripes on the rectangle.

3 Put dots on the hexagon.

4 Colour the pentagon red.

5 Colour the octagon blue.

# Number and Algebra

## SET 3 Counting patterns with fractions

Write the next 2 mixed numerals in each sequence.

**1** | $0$ | $\frac{1}{4}$ | $\frac{2}{4}$ | $\frac{3}{4}$ | $1$ | | |

**2** | $0$ | $\frac{1}{2}$ | $1$ | $1\frac{1}{2}$ | $2$ | $2\frac{1}{2}$ | | |

**3** | $0$ | $\frac{1}{5}$ | $\frac{2}{5}$ | $\frac{3}{5}$ | $\frac{4}{5}$ | $1$ | | |

**4** | $0$ | $\frac{1}{8}$ | $\frac{2}{8}$ | $\frac{3}{8}$ | $\frac{4}{8}$ | $\frac{5}{8}$ | $\frac{6}{8}$ | $\frac{7}{8}$ | $1$ | | |

Place these mixed numerals in ascending order.

**5** $\frac{3}{10}, \frac{1}{10}, \frac{7}{10}$ ______________

**6** $\frac{3}{4}, \frac{1}{4}, \frac{2}{4}$ ______________

**7** $\frac{7}{8}, \frac{3}{8}, \frac{5}{8}$ ______________

**8** $1\frac{1}{4}, 1\frac{3}{4}, 1\frac{2}{4}$ ______________

**9** $2\frac{3}{5}, 2\frac{1}{5}, 2\frac{2}{5}$ ______________

True or false?

**10** $1\frac{7}{10} < 1\frac{9}{10}$ ______________

**11** $2\frac{1}{4} > 2\frac{3}{4}$ ______________

**12** $2\frac{1}{2} > 2\frac{1}{4}$ ______________

**13** $2\frac{1}{8} > 2\frac{1}{4}$ ______________

## SET 4 Extension

**1** Add \$6 to \$17.80.

**2** Subtract \$9 from \$13.60.

**3** 5 pens at \$1.95 each

**4** 19 places behind 35th

**5** What number is halfway between 38 and 46?

**6** 18, 36, ☐, ☐, 90, ☐

**7** If 4 rulers cost 88c, how much would 6 cost?

**8** Change from \$5 if I spent \$3.45

**9** List the factors of 36.

**10** How many centimetres in $3\frac{1}{2}$ m?

**11**

| Tathra | | Green Point | |
|---|---|---|---|
| Points | 69 | Points | 95 |

Which team won and what was the difference in their point scores?

**12** Round off to the nearest 10 to find an answer for 427 + 153.

**13** ☐ × (5 + 6) = 99

# Measurement Millilitres

Find the differences in capacity between these items.

**1** A can of soft drink and a small milk carton. ☐

**2** A yoghurt and a medicine bottle. ☐

**3** A carton of cream and a Fizzo bottle. ☐

**4** A large milk carton and a carton of cream. ☐

**5** A soft drink can and a yoghurt tub. ☐

**6** A Fizzo bottle and a medicine bottle. ☐

Yoghurt 100mL

# Number and Algebra

## SET 1 Basic

**1** 47 − 5

**2** 42 − 5

**3** 26 + 7

**4** 36 + 7

**5** 7 × 4

**6** 6 × 3

**7** 5 × 7

**8** 8 × 6

**9** 7 + 3 + 12

**10** 8 + 7 − 5

**11** What is the difference between 20 and 6?

**12** How many fours in 16?

**13** What is the sum of 52 and 11?

**14** What is the product of 9 and 2?

**15** How much will it cost Tran to buy her lunch?

| Tran | Cost |
|---|---|
| 1 Sandwich | $1.20 |
| 1 Drink | $0.80 |
| 1 Apple | $0.60 |
| Total | |

## SET 2 Subtraction to 9999

**1**

| | THOU | HUND | TENS | ONES |
|---|---|---|---|---|
| | 8 | 7 | 6 | 4 |
| − | 3 | 4 | 5 | 2 |
| | | | | |

**2**

| | THOU | HUND | TENS | ONES |
|---|---|---|---|---|
| | 6 | 7 | 4 | 2 |
| − | 4 | 3 | 1 | 8 |
| | | | | |

**3**

| | THOU | HUND | TENS | ONES |
|---|---|---|---|---|
| | 8 | 8 | 2 | 5 |
| − | 5 | 3 | 4 | 6 |
| | | | | |

**4**

| | THOU | HUND | TENS | ONES |
|---|---|---|---|---|
| | 7 | 3 | 6 | 2 |
| − | 2 | 4 | 5 | 5 |
| | | | | |

Mathematical Reasoning

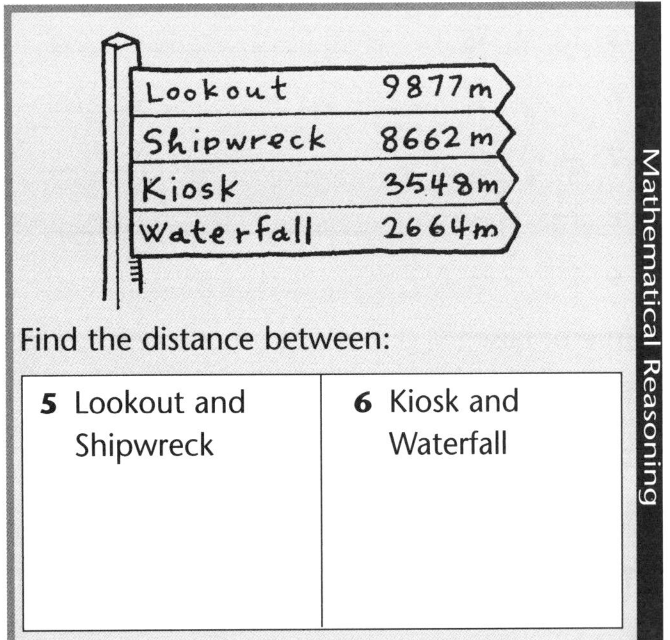


Find the distance between:

| **5** Lookout and Shipwreck | **6** Kiosk and Waterfall |
|---|---|
| | |

# Statistics and Probability Chance

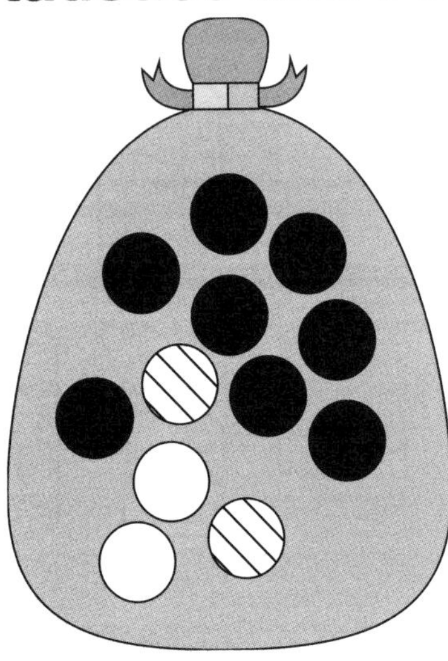

Mike is blindfolded and pulls out marbles from his bag. Use the words likely, unlikely or impossible to describe the chances of him pulling out

**1** a white marble ____________

**2** a striped marble ____________

**3** a black marble ____________

**4** a dotted marble ____________

# Number and Algebra

## SET 3 Odd and even numbers

Answer the questions.

1 If you add two odd numbers can you get a total which is an odd number? ______

2 If you add two even numbers can you get a total which is an odd number? ______

3 If you add an even number and an odd number can you get a total which is an odd number? ______

4 If you take away an odd number from another odd number can you get an answer which is an odd number?

______

5 If you multiply two even numbers can you get a product which is an odd number? ______

6 If you multiply an even number by an odd number can you get a product which is an odd number? ______

7 If you multiply two odd numbers can you get a product which is an odd number? ______

## SET 4 Extension

1 3 m and 17 cm = ☐ cm

2 274, 271, 268, ☐

3 How many centimetres in $4\frac{1}{2}$ m?

4 147, 151, 155, ☐

5 Value of 6 in 7631

6 $\frac{1}{4}$ of $72

7 What is 17 more than 191?

8 What is the sum of 237 and 122?

9 Halve 380.

10 How many sides on 3 octagons?

11 How much is $4\frac{1}{2}$ kg of meat at $12 per kilogram?

12 12 ÷ 3 + 37

13 How many tens in 376?

14 How many halves in 4 wholes?

15 Write 4207 in words.

______

______

16 List the factors of 48.

17 Write $\frac{3}{10}$ as a decimal.

18 Round 3866 to the nearest 100.

# Measurement Grams

Record these measurements in abbreviated form.

1 1 kilogram and 600 grams = 1 kg 600 g

2 2 kilograms and 260 grams = ______

3 4 kilograms and 950 grams = ______

4 5 kilograms and 20 grams = ______

5 3 kilograms and 35 grams = ______

# Number and Algebra

## SET 1 Basic

1 40 + 7

2 83 + 6

3 27 − 5

4 45c − 4c

5 7 × 6

6 7 × 3

7 7 × 4

8 7 × 5

9 24 take away 5

10 8 + 9 − 5

11 (7 + 3) × 2

12 \$8.93 = ☐ c

13 How many sixes in 18?

14 What is the product of 8 and 8?

15

## SET 2 Improper fractions

Write the improper fraction in the answer boxes.

1 $1\frac{1}{2}$

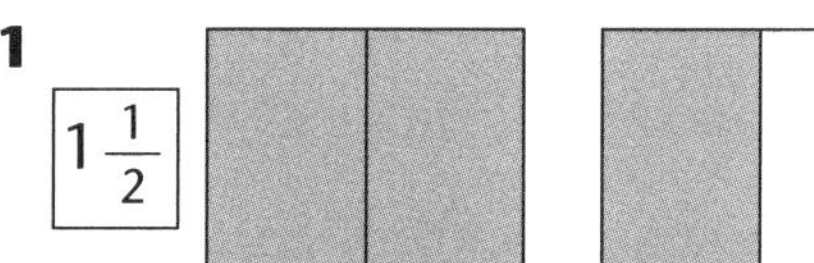

2 $1\frac{1}{4}$

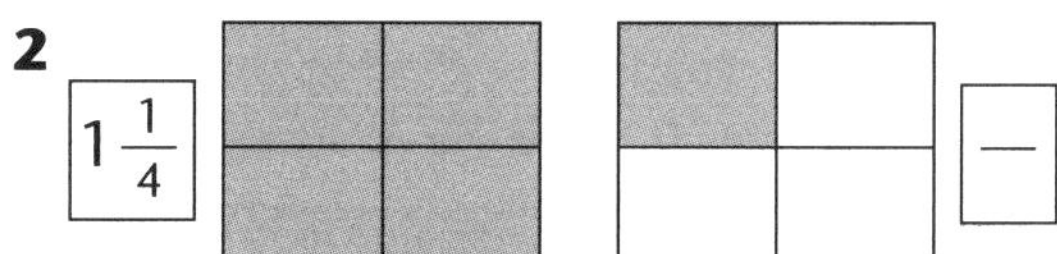

3 $1\frac{1}{3}$

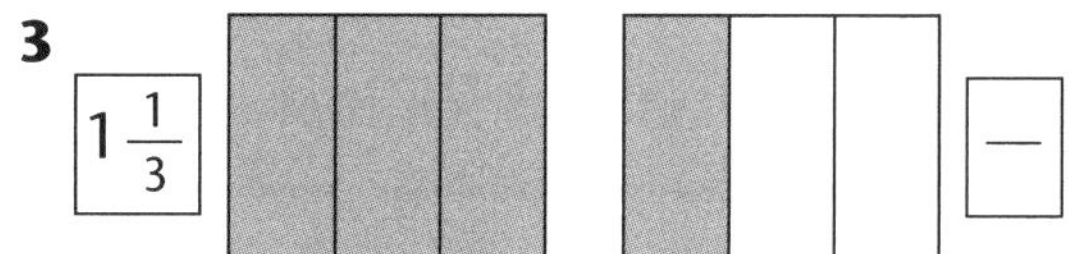

4 $1\frac{3}{4}$

5 $1\frac{3}{5}$

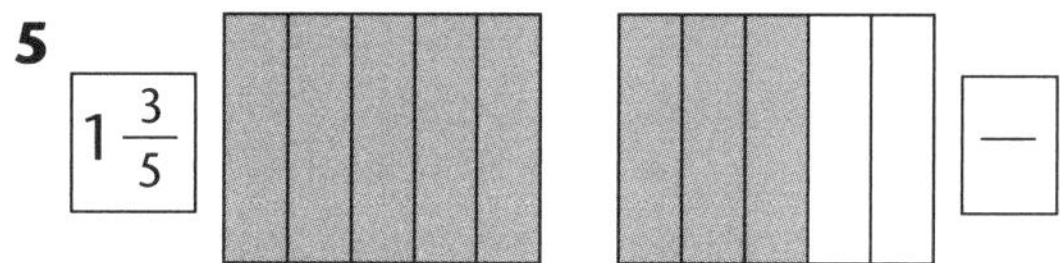

6 $1\frac{5}{8}$

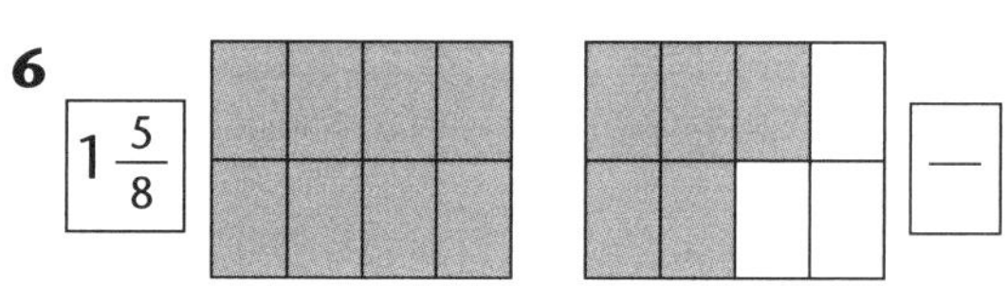

# Space Symmetrical patterns

Complete the symmetrical pattern.

# Number and Algebra

## SET 3 Number relationships

Complete the number patterns.

1

| Hours | 1 | 2 | 3 | 4 | 5 | 6 |
|---|---|---|---|---|---|---|
| Pay | $8 | $16 | $24 | | | |

2

| Hours | 1 | 2 | 3 | 4 | 5 | 6 |
|---|---|---|---|---|---|---|
| Kilometers | 4 | 8 | 12 | | | |

3

| Hours | 1 | 2 | 3 | 4 | 5 | 6 |
|---|---|---|---|---|---|---|
| Pay | $7 | $14 | $21 | | | |

4

| Hours | 1 | 2 | 3 | 4 | 5 | 6 |
|---|---|---|---|---|---|---|
| Kilograms | 9 | 18 | 27 | | | |

5

| Kilograms | 1 | 2 | 3 | 4 | 5 | 6 |
|---|---|---|---|---|---|---|
| Tomatoes | 6 | 12 | 18 | | | |

## SET 4 Extension

**1** (4 + 5) x 4

**2** 100 + 60 + 97

**3** Estimate an answer to 399 + 148.

**4** $\frac{1}{4}$ of 40 marbles

**5** Add $4 to $10.95.

**6** Half of $6.20

**7** 1323 + 501

**8** Subtract $15 from $48.50.

**9** 70 + 70 + 70 + 70 + 70

**10** Half of $8.60

**11** 1027, 1127, 1227, ☐, ☐

**12** If 3 kg cost $6, how much would 7 kg cost?

**13** How many legs on 2 dogs, 4 spiders and 2 children?

**14** Cost of 3 books at $43 each

**15** Write 2495 in words. ____________________

____________________________________

**16** Write the largest number you can using 7, 0, 5, 9.

**17** Is a gluestick a cone?

# Measurement Analog clocks

Draw these times on the clock faces.

**1** 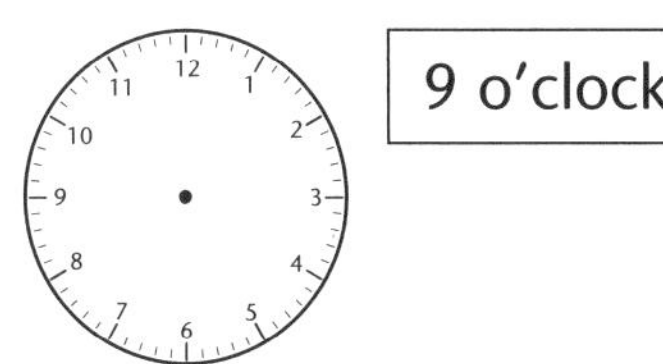
9 o'clock

**2** 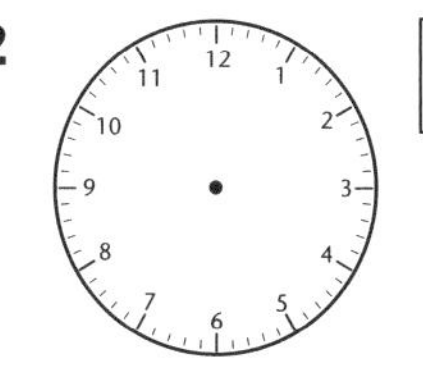
$\frac{1}{2}$ past 3

**3** 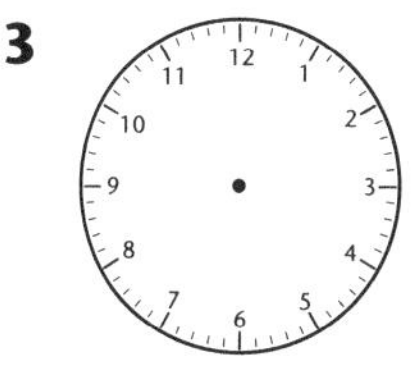
25 past 7

**4** 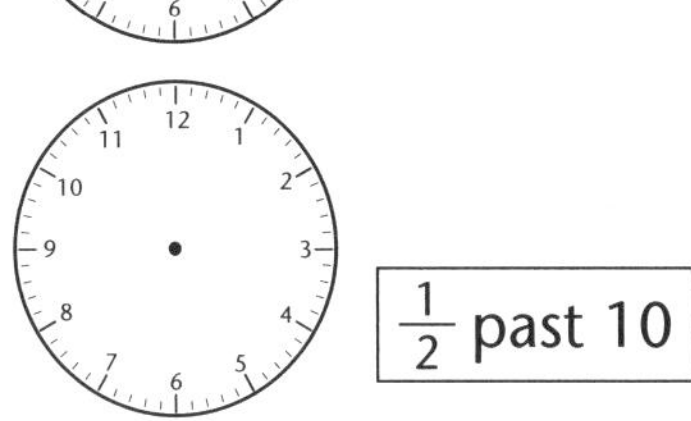
$\frac{1}{2}$ past 10

**5** 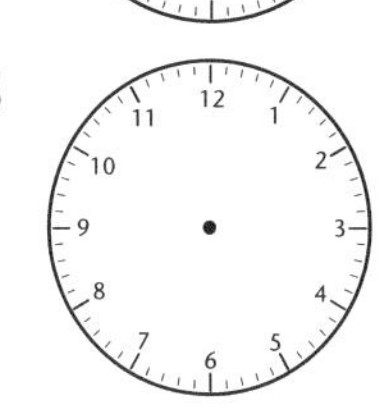
$\frac{1}{4}$ past 6

**6** 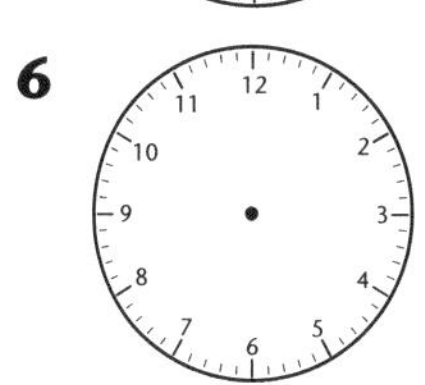
10 to 9

# Number and Algebra

## SET 1 Basic

**1** 9 + 5

**2** 5 × 7

**3** 5 × 4 − 1

**4** $7 + $7

**5** $4^2$

**6** 5 + 4 + 3 + 2

**7** Seven at 10c each

**8** 94 less 5

**9** 73 kg − 10 kg

**10** What is the product of 8 and 9?

**11** What is the difference between 35 and 6?

**12** Halve 10 × 4.

**13** Subtract 7 from 21.

**14** 962 = ☐ hundreds + ☐ tens + ☐ ones.

**15**

The 3:12 train from the city was 10 minutes late. When did it arrive?

☐

## SET 2 Addition and subtraction

Solve the additions and subtractions.

**1**

| | THOU | HUND | TENS | ONES |
|---|---|---|---|---|
| | 3 | 7 | 8 | 7 |
| + | 6 | 0 | 5 | 4 |
| | | | | |

**2**

| | THOU | HUND | TENS | ONES |
|---|---|---|---|---|
| | 3 | 8 | 7 | 6 |
| + | 5 | 8 | 5 | 6 |
| | | | | |

**3**

| | THOU | HUND | TENS | ONES |
|---|---|---|---|---|
| | 8 | 6 | 9 | 7 |
| − | 3 | 5 | 2 | 8 |
| | | | | |

**4**

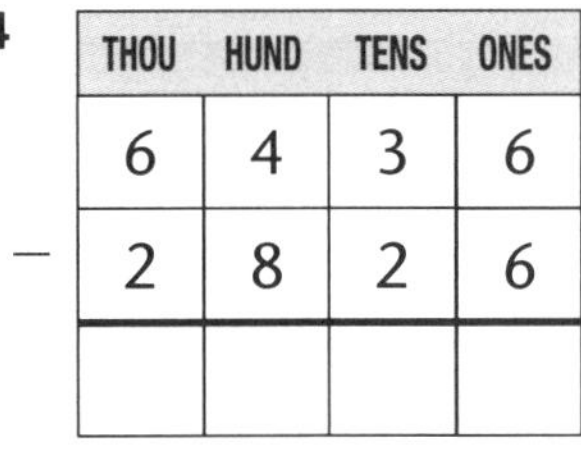

| | THOU | HUND | TENS | ONES |
|---|---|---|---|---|
| | 6 | 4 | 3 | 6 |
| − | 2 | 8 | 2 | 6 |
| | | | | |

**5** Jack saved $1346 but needed $3524 to buy a new computer. How much more does he need to save?

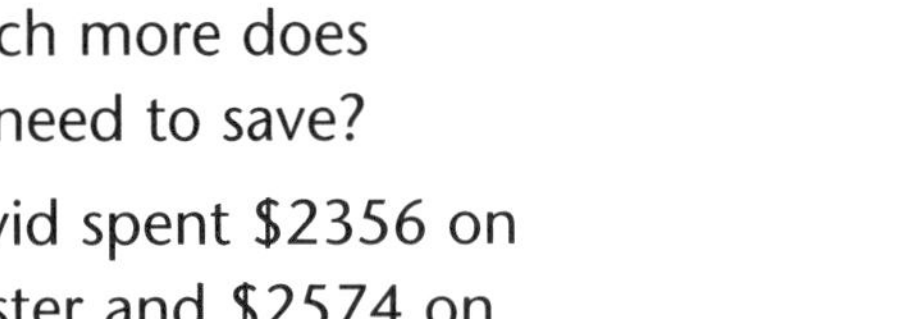

**6** David spent $2356 on plaster and $2574 on tradesmen to put it on. How much did he spend altogether?

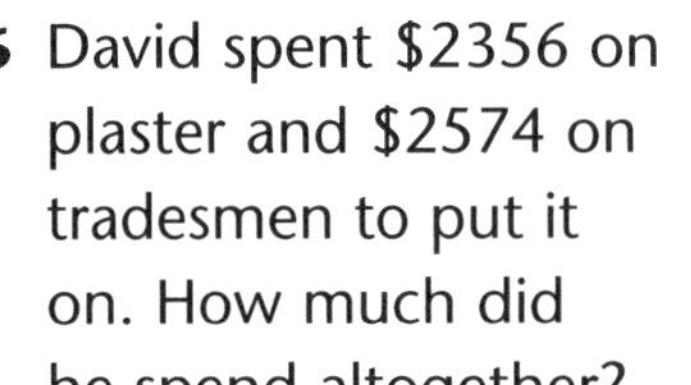

**7** Jenny saved $1213 each month for 3 months. What were her total savings?

# Measurement Informal volume

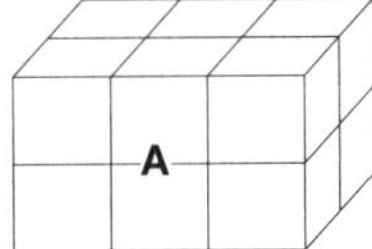

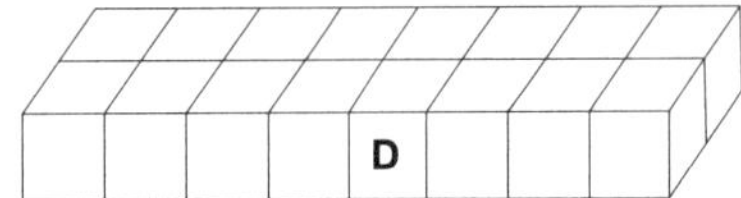

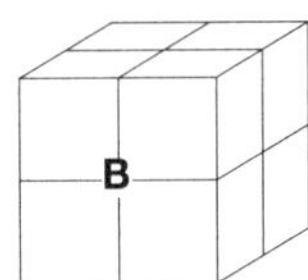

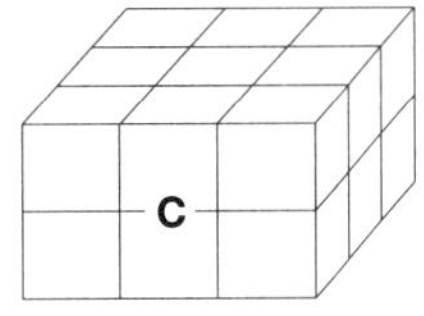



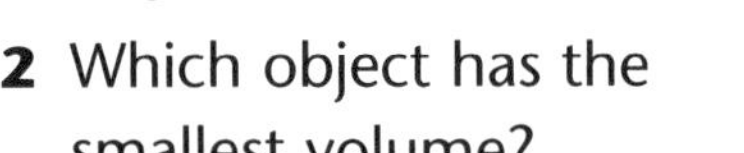

**1** Which object has the largest volume? ________

**2** Which object has the smallest volume? ________

**3** Which object has a volume of 16 cubes? ________

# Number and Algebra

## SET 3 Patterns in tables

Follow the rules to complete the number patterns.

1

| Add 5 | | | | | | | |
|---|---|---|---|---|---|---|---|
| 1 | 2 | 3 | 4 | 5 | 6 | 7 | 8 |
| | | | | | | | |

2

| Add 9 | | | | | | | |
|---|---|---|---|---|---|---|---|
| 1 | 2 | 3 | 4 | 5 | 6 | 7 | 8 |
| | | | | | | | |

3

| Multiply by 4 | | | | | | | |
|---|---|---|---|---|---|---|---|
| 1 | 2 | 3 | 4 | 5 | 6 | 7 | 8 |
| | | | | | | | |

Complete the tables to answer these problems.

Mathematical Reasoning

4 Tom walks at a rate of 5 km per hour. How far does he walk in 7 hours?

| Hours | 1 | 2 | 3 | 4 | 5 | 6 | 7 |
|---|---|---|---|---|---|---|---|
| Km | | | | | | | |

5 Cans of soup are packed in boxes of 8. How many cans in 7 boxes?

| Boxes | 1 | 2 | 3 | 4 | 5 | 6 | 7 |
|---|---|---|---|---|---|---|---|
| Cans | | | | | | | |

## SET 4 Extension

1 2 m and 37 cm = ☐ cm

2 How many millilitres in 4 L?

3 $36 \div 6$

4 $36 \div 4 + 5$

5 4976, 4986, ☐, 5006

6 $3 \times 10 + 9$

7 How much are 3 cakes at $13 each?

8 Write 65 in Roman numerals.

9 What number is halfway between 1350 and 1450?

10 $\frac{1}{5}$ of 25

11 How many angles on 7 hexagons?

Mathematical Reasoning

12 How many red pegs are there if there are 24 pegs on the clothes line? ________

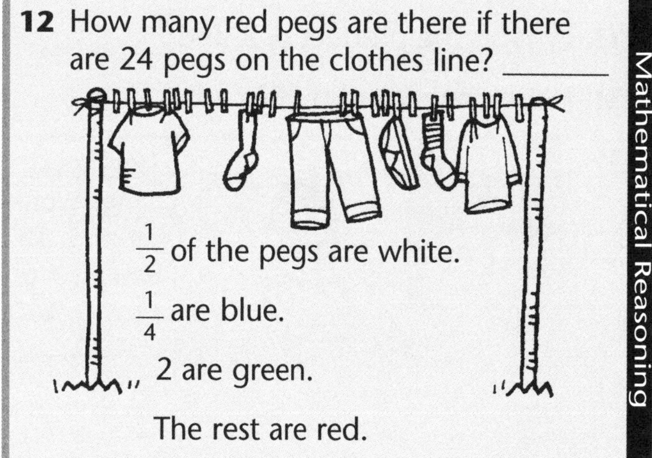

$\frac{1}{2}$ of the pegs are white.

$\frac{1}{4}$ are blue.

2 are green.

The rest are red.

# Statistics and Probability Chance

1 Colour the label that best describes the chance of 2 dice landing on ones.

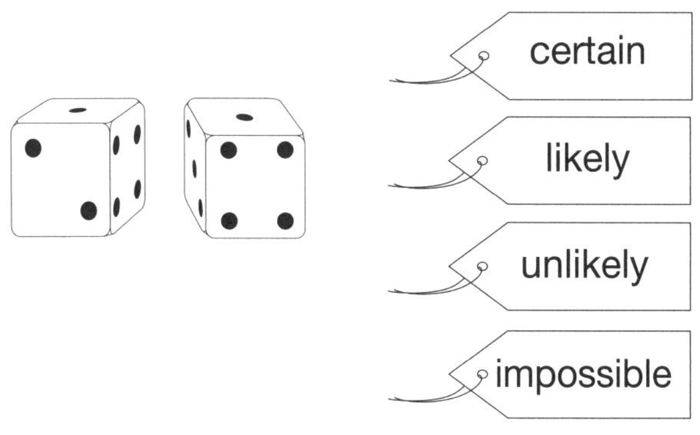


2 Colour the label that describes the chance of the spinner landing on red.

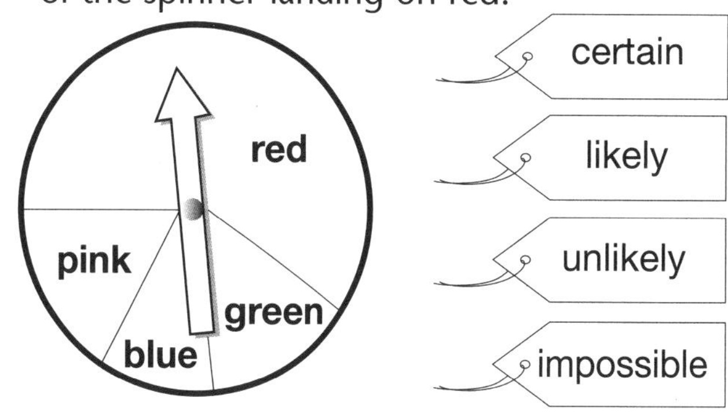

# Number and Algebra

## SET 1 Basic

1 9 + 7

2 8 + 5

3 7 × 6

4 7 × 4

5 7 × 8

6 7 × 7

7 68 – 6

8 52 – 6

9 8 + 8 + 8 – 4

10 7 + 10 – 3 + 4

11 \$5.86 = ☐ c

12 How many 20c coins make \$2.60?

13 What is the product of 8 and 7?

14 How many fours in 28?

15

## SET 2 Inverse operations

Write an inverse operation to check each number sentence.

1 19 + 6 = 25 | 25 | – | ☐ | = | ☐

2 25 – 12 = 13 | 13 | + | ☐ | = | ☐

3 7 × 6 = 42 | 42 | ÷ | ☐ | = | ☐

4 45 ÷ 5 = 9 | 9 | × | ☐ | = | ☐

5 36 ÷ 6 = 6 | 6 | × | ☐ | = | ☐

Write a subtraction to check each addition.

6

| | THOU | HUND | TENS | ONES |
|---|---|---|---|---|
| | 8 | 7 | 6 | 8 |
| + | | 4 | 2 | 5 |
| | 9 | 1 | 9 | 3 |

| | THOU | HUND | TENS | ONES |
|---|---|---|---|---|
| | 9 | 1 | 9 | 3 |
| – | | | | |
| | | | | |

7

| | THOU | HUND | TENS | ONES |
|---|---|---|---|---|
| | 5 | 6 | 8 | 8 |
| + | 2 | 3 | 4 | 5 |
| | 8 | 0 | 3 | 3 |

| | THOU | HUND | TENS | ONES |
|---|---|---|---|---|
| | 8 | 0 | 3 | 3 |
| – | | | | |
| | | | | |

# Measurement Temperature

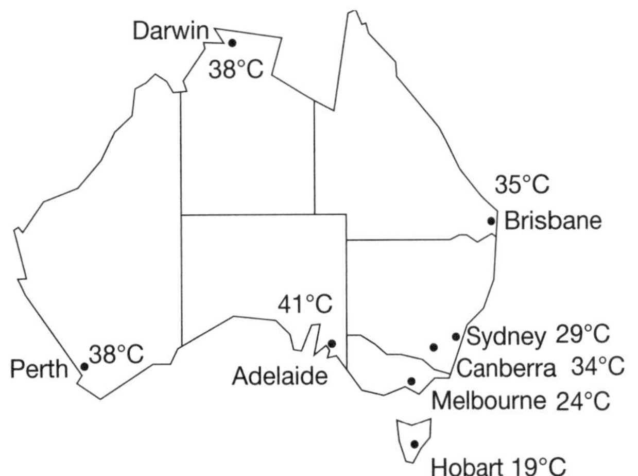


What is the difference in temperature between:

1 Darwin and Sydney? ________

2 Melbourne and Hobart? ________

3 Canberra and Perth? ________

4 Perth and Melbourne? ________

5 Brisbane and Canberra? ________

6 Adelaide and Hobart? ________

# Number and Algebra

## SET 3 Decimals

Shade the hundredths grids to represent the decimals.

1 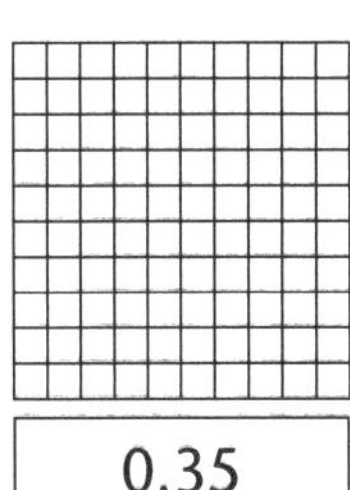

0.35

2 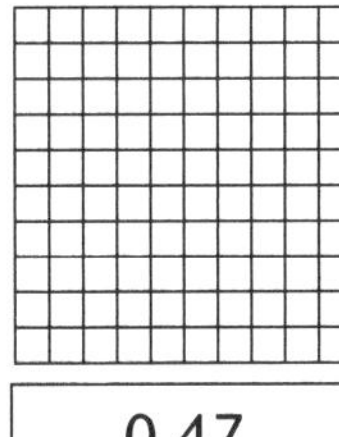

0.47

3 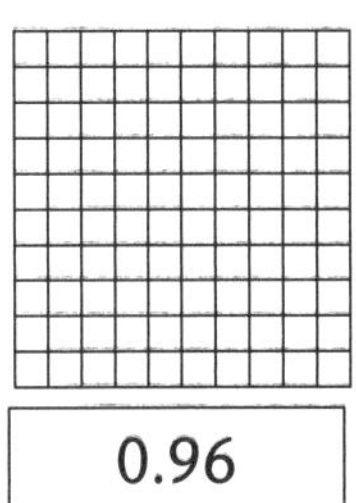

0.96

4 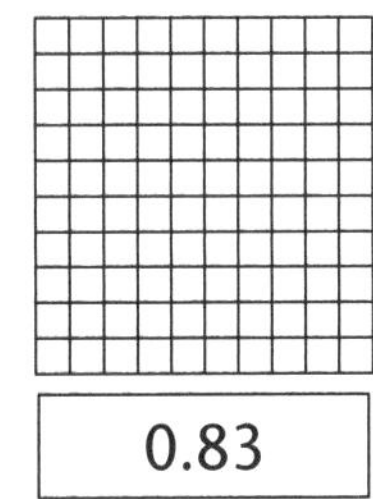

0.83

5 Write $\frac{27}{100}$ as a decimal.

6 Write $\frac{54}{100}$ as a decimal.

Order the decimals from smallest to largest.

7 0.35, 0.29, 0.09 ____________

8 0.33, 0.39, 0.29 ____________

9 0.87, 1.07, 7.01 ____________

10 9.85, 8.59, 9.58 ____________

## SET 4 Extension

1 6 tens + 423

2 437 cm = ☐ m and ☐ cm

3 36 ÷ 6 × 3

4 How many grams in $3\frac{1}{2}$ kg?

5 120 minutes = ☐ hours.

6 What number represents thousands in 7523?

7 Share $36 among 6 people.

8 8846, 8746, 8646, ☐

9 How many hundreds in 4231?

10 How many legs on 9 beetles?

11 Estimate an answer to 137 + 249.

12 3796 + 203

13 If 10 apples cost $1, how much for 50 apples?

**Mathematical Reasoning**

14 The perimeter of a rectangle is 120 cm. What are its dimensions if its length is twice its width?

Length

Width

Length ____________

Width ____________

# Measurement Square centimetres

1 Draw a square with an area of 16 $cm^2$.

2 Draw a rectangle with an area of 16 $cm^2$.

# Number and Algebra

## SET 1 Basic

1 6 + 2 + 8

2 9 + 9

3 16 – 6

4 5 × 6 – 1

5 6c + 4c + 11c

6 25 take away 4

7 $9.27 = ☐ c

8 How many sixes in 18?

9 Subtract 11 from 40.

10 7 tens × 6

11 Share $42 among 6 people.

12 Eight at $11 each

13 65 kg – 9 kg

14 Write 2060 in words.

______________________________

______________________________

## SET 2 Division

Answer the divisions.

1 12 ÷ 2 = ☐

2 12 ÷ 3 = ☐

3 16 ÷ 4 = ☐

4 25 ÷ 5 = ☐

5 30 ÷ 6 = ☐

6 49 ÷ 7 = ☐

7 64 ÷ 8 = ☐

8 72 ÷ 9 = ☐

9 Farmer Tom shared 45 sheep into 5 paddocks. How many sheep were in each paddock?

10 Sita had 48 stamps that she put into 4 albums. How many stamps did she put in each album?

**Mathematical Reasoning**

Write a division question for the answers.

11 ☐ ÷ ☐ = 5

12 ☐ ÷ ☐ = 7

# Statistics and Probability Survey

**Most popular snacks of Ms Johnson's class**

| Snack | Tally | Number |
|---|---|---|
| Cakes | 卌 卌 \| | |
| Biscuits | \|\|\| | |
| Pancakes | 卌 卌 | |
| Popcorn | 卌 \|\| | |

Construct a column graph from the data.

1

**Popular snacks**

| | | | | | | | | | | | | |
|---|---|---|---|---|---|---|---|---|---|---|---|---|
| Cakes | | | | | | | | | | | | |
| Biscuits | | | | | | | | | | | | |
| Pancakes | | | | | | | | | | | | |
| Popcorn | | | | | | | | | | | | |

**Number**

2 Which was the most popular food?

3 How many children are in Ms Johnson's class?

# Number and Algebra

## SET 3 Tenths as decimals

Write each tenth as a decimal.

1 $\frac{1}{10}$

2 $\frac{7}{10}$

3 $\frac{9}{10}$

4 $\frac{3}{10}$

5 $\frac{5}{10}$

6 $\frac{8}{10}$

7 $\frac{4}{10}$

8 $\frac{6}{10}$

Shade each shape to match the decimal label.

9 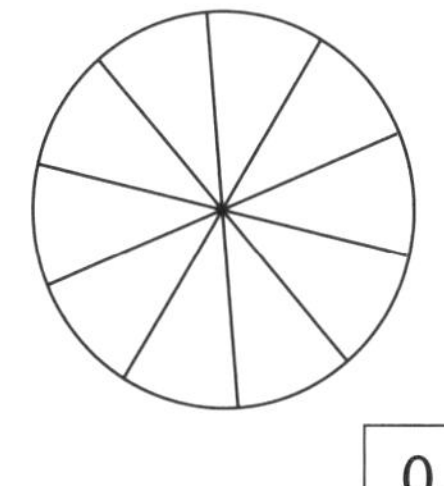

0.7

10 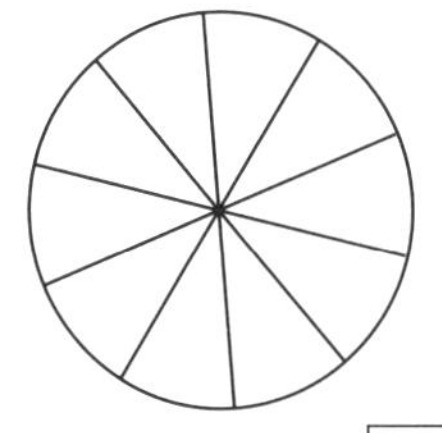

0.1

11 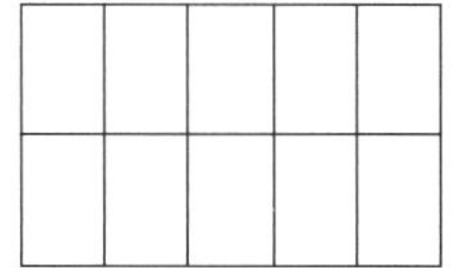

0.9

12 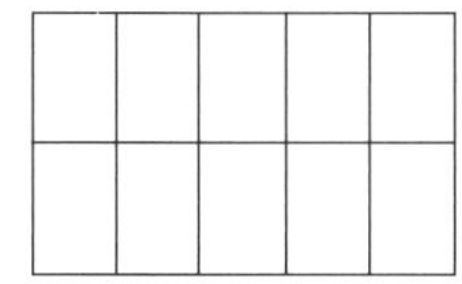

0.6

13 Complete the pattern.

| 0.1 | 0.3 | 0.5 | | | |
|---|---|---|---|---|---|

## SET 4 Extension

1 (2 + 6) × 4

2 1200 + 50 + 8

3 How many sixes in 48?

4 How many cents in $17.06?

5 Add $5 to $19.20.

6 Half of $6.10

7 (3 + 7) × 5

8 Subtract $9 from $19.95.

9 20 + 20 + 20 + 20 + 20 + 20 + 20

10 $\frac{1}{4}$ of $4.68

11 How much is 3 kg at $2.50 per kg?

12 If 3 kg cost $33, how much would 7 kg cost?

13 13 less than 9963

14 How many metres in 700 cm?

15 What is the perimeter of a square with 7 cm sides?

**Mathematical Reasoning**

16 How old is Annabella if she is $\frac{1}{4}$ the age of Susan who is 28?

# Measurement Perimeter

Calculate the perimeter of each shape.

1 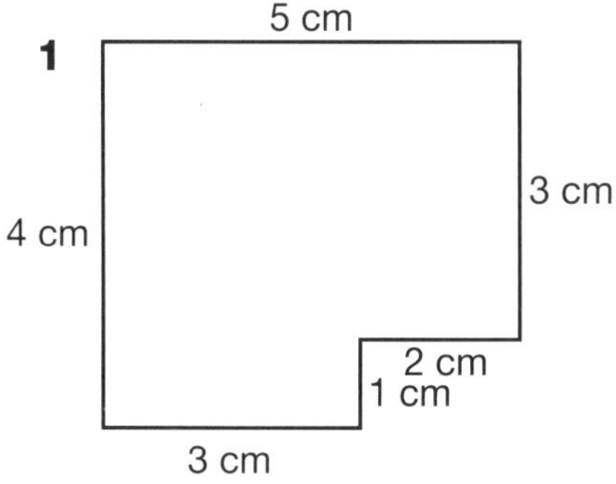


Perimeter = ________ cm

2 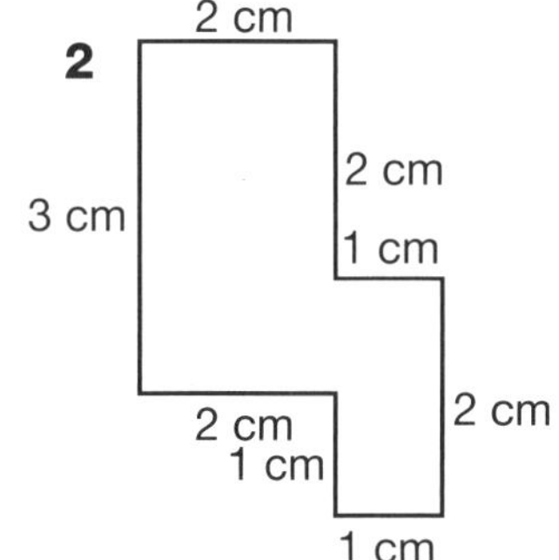


Perimeter = ________ cm

3 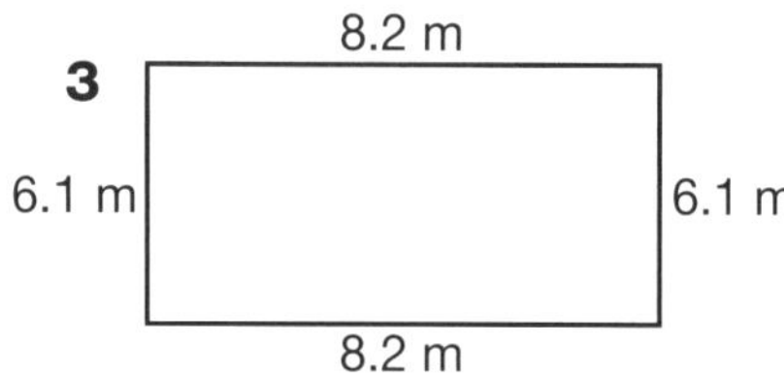


Perimeter = ________ m

UNIT 21

## Number and Algebra

### SET 1 Basic

1 18 – 8

2 18 – 10

3 12 + 6

4 18 + 18

5 7 × 2

6 6 × 8

7 4 × 9

8 7 × 3

9 6 + 7 + 6 + 2

10 23 take away 6

11 370c = $ ☐

12 How many fours in 24?

13 What is the sum of 19 and 6?

14 How many 20c coins in $1?

15 4057 = ☐ thousands + ☐ hundreds + ☐ tens + ☐ ones

### SET 2 Division strategies

Solve the divisions then write a multiplication fact from each division fact.

1 25 ÷ 5 = 5    5 × 5 = 25

2 32 ÷ 8 = ☐    ☐ × ☐ = ☐

3 40 ÷ 10 = ☐    ☐ × ☐ = ☐

4 28 ÷ 7 = ☐    ☐ × ☐ = ☐

5 81 ÷ 9 = ☐    ☐ × ☐ = ☐

6 42 ÷ 7 = ☐    ☐ × ☐ = ☐

7 36 ÷ 6 = ☐    ☐ × ☐ = ☐

8 54 ÷ 6 = ☐    ☐ × ☐ = ☐

9 Trent shared 29 cakes between 3 boys and himself. How many did each boy receive?

☐ remainder ☐

10 How many groups of 7 can be made from 23 people?

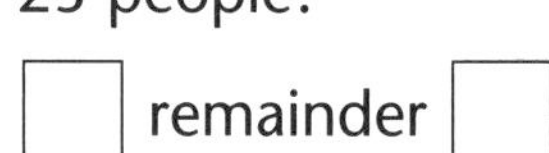

## Space Triangles

Draw a line to match the triangle to its label.

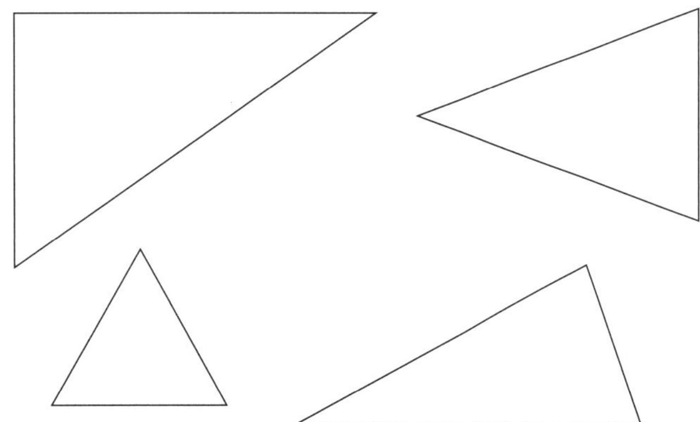

Equilateral triangle

Scalene triangle

Isosceles triangle

Right angle triangle

# Number and Algebra

## SET 3 Tenths and hundredths

Complete the table.

| | Tenths | Hundredths | Decimal |
|---|---|---|---|
| 1 | $\frac{1}{10}$ | | 0.10 |
| 2 | $\frac{3}{10}$ | | 0.30 |
| 3 | | $\frac{90}{100}$ | 0.90 |
| 4 | $\frac{5}{10}$ | $\frac{50}{100}$ | 0.50 |
| 5 | $\frac{7}{10}$ | | 0.70 |
| 6 | | $\frac{20}{100}$ | 0.20 |
| 7 | $\frac{6}{10}$ | $\frac{60}{100}$ | |

Write *true* or *false*.

**8** $\frac{3}{10} = 0.3$ ______

**9** $\frac{7}{10} = \frac{70}{100}$ ______

**10** $0.5 = \frac{5}{10}$ ______

**11** $\frac{2}{10} > \frac{5}{10}$ ______

**12** $0.3 < 0.7$ ______

**13** $\frac{30}{100} = \frac{3}{10}$ ______

**14** $0.6 < 0.2$ ______

**15** $\frac{5}{10} > \frac{8}{10}$ ______

**16** $\frac{40}{100} < 0.5$ ______

## SET 4 Extension

**1** How many grams in 3 kg?

**2** 37, 137, 237, ☐

**3** 4 m and 5 cm = ☐ cm

**4** $30 \div 5 \times 6$

**5** What is the value of 1 in 4107?

**6** How many 25 cm lengths in 3 m?

**7** What number is halfway between 126 and 130?

**8** 1236 + 603

**9** $\frac{1}{5}$ of $80

**10** How many days in winter?

**11** $\frac{1}{2}$ kg meat at $11 per kilogram.

**12** How much is left from $5, if I spent $1.75?

**13** Share 37 among 5.

**Mathematical Reasoning**

**14** Heather was supposed to cut a 180 cm piece of rope in half but she made a mistake and one piece was 20 cm longer than the other. How long are her pieces of rope?

0 cm 180 cm

# Measurement am and pm time

Write the digital time for each clock time face using am and pm notation.

**1** 5:00 am | morning

**2** 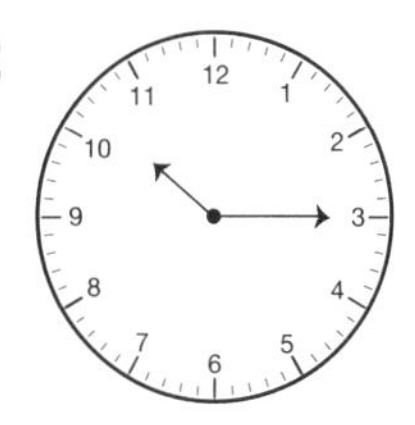 : | evening

**3** 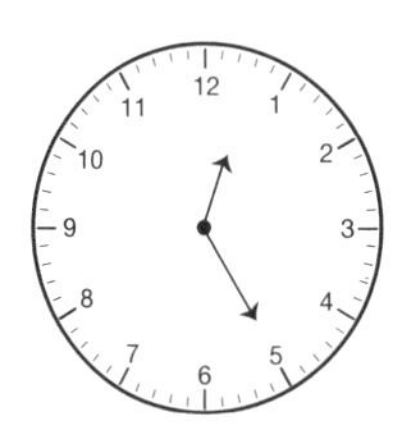 : | afternoon

**4** 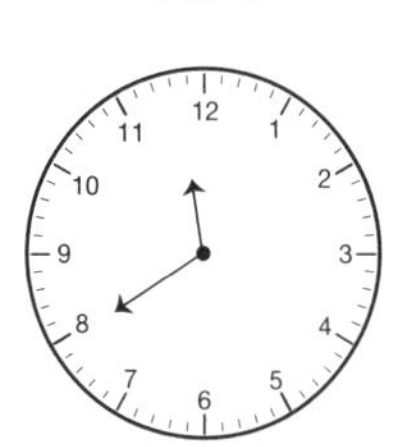 : | evening

**5** 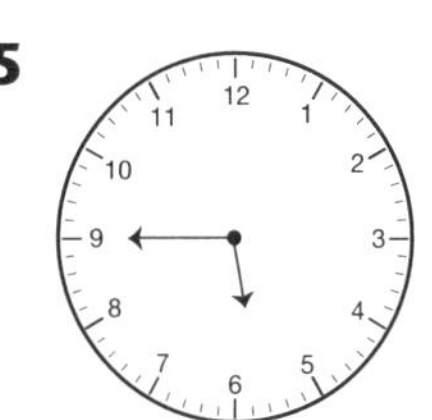 : | morning

**6** 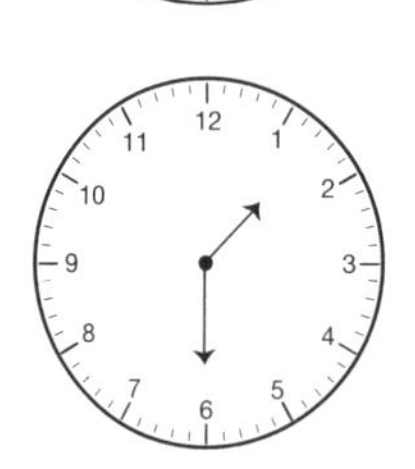 : | afternoon

# Number and Algebra

## SET 1 Basic

**1** 5 × 7

**2** 20 – 5

**3** 9 ÷ 3

**4** 7 × 5

**5** 13 – 5

**6** 7 + 9 – 6

**7** What is the sum of 7 and 23?

**8** What is the product of 8 and 3?

**9** How many eights in 24?

**10** 95, 100, ☐, 110

**11** A quarter of 12

**12** \$4.07 = ☐ c

**13** What is the difference between 20 and 6?

**14** 3204 = ☐ thousands + ☐ hundreds + ☐ tens + ☐ ones

**15**

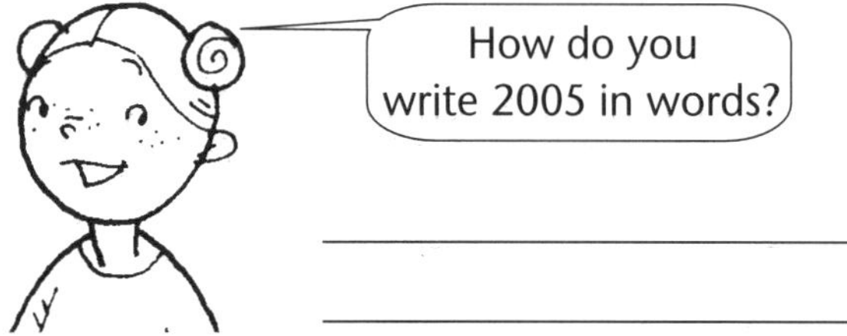


______________________

______________________

## SET 2 Decimal place value

What is the place value of each bold number?

**1** 1.**2**3 ______________

**2** 5.3**4** ______________

**3** **1**3.35 ______________

**4** 1**7**.56 ______________

**5** 3.4**7** ______________

**6** 13.**6**3 ______________

**7** **3**67.5 ______________

**8** 37.**5**3 ______________

Order these decimals from smallest to largest.

**9**

| 0.67 | 0.76 | 0.53 |
|---|---|---|
| | | |

**10**

| 1.07 | 1.86 | 0.93 |
|---|---|---|
| | | |

**11**

| 12.03 | 1.24 | 1.42 |
|---|---|---|
| | | |

# Space Using a scale

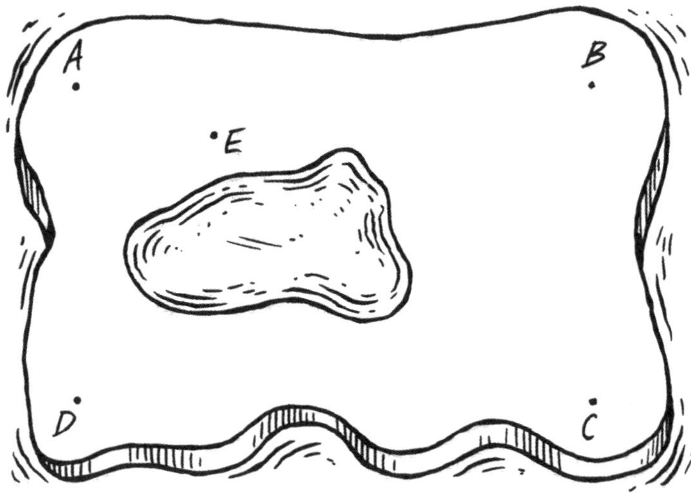


Scale 1 cm = 10 m

Using the scale calculate the distance to the nearest 10 metres between:

**a** A and B ________ m

**b** B and C ________ m

**c** C and D ________ m

**d** E and A ________ m

# Number and Algebra

## SET 3 Division strategies

Solve the divisions.

| | | |
|---|---|---|
| **1** 2)36 | **2** 3)51 | **3** 2)52 |
| **4** 3)48 | **5** 3)54 | **6** 4)60 |
| **7** 5)65 | **8** 5)85 | **9** 6)78 |
| **10** 6)84 | **11** 7)84 | **12** 4)72 |

**13** 6 boys shared 36 football cards. How many cards did each boy receive?

**14** 75 children were put into 5 equal groups. How many children in each group?

**15** 64 cakes were put onto 4 trays. How many cakes on each tray?

## SET 4 Extension

**1** How many 50 cm lengths in 5 m?

**2** 18 – 5 + 6

**3** 6 kg at $5.50 per kg

**4** Subtract $8 from $2479.

**5** Share 56 marbles among 7 children.

**6** How many sides have 8 octagons?

**7** How many 250 g bags in a kilogram?

**8** 4435 + 402

**9** What is the value of 5 in 7850?

**10** Three positions after 429th

**11** How much is left from $10 if I spent $1.65?

**12** Half of 286

**13** 45 ÷ 3

**14** How many tens in 489?

**15** Write 4309 in words. ______________________

______________________________

**16** How much did Liam spend at the canteen if he bought the following?

| | | |
|---|---|---|
| 1 | SANDWICH | $1.50 |
| 1 | SAUSAGE ROLL | $0.80 |
| 1 | ORANGE | $0.45 |
| 1 | FROZEN YOGHURT | $0.85 |

# Measurement Volume

Record the volume of each object in cubes.

**1**

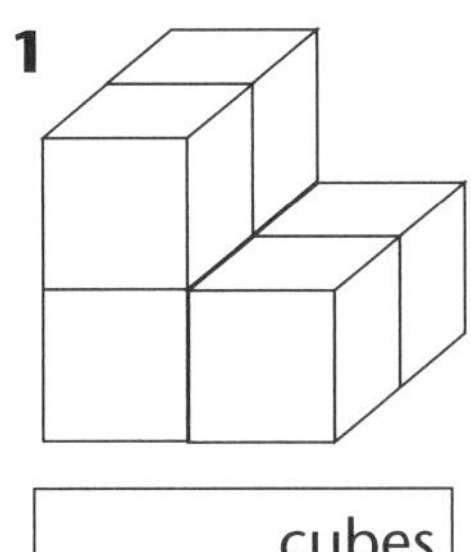

[ ] cubes

**2**

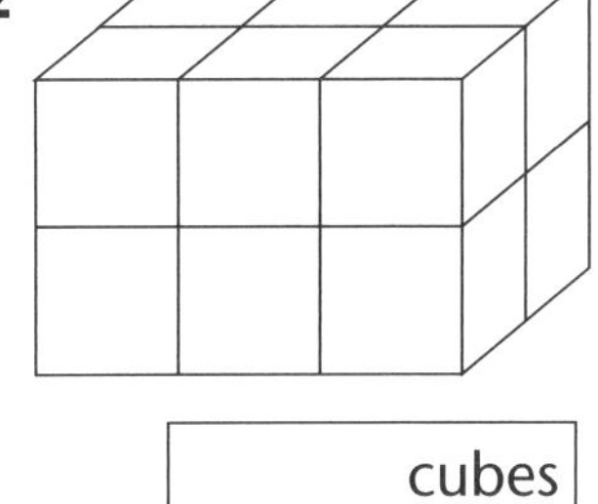

[ ] cubes

**3**

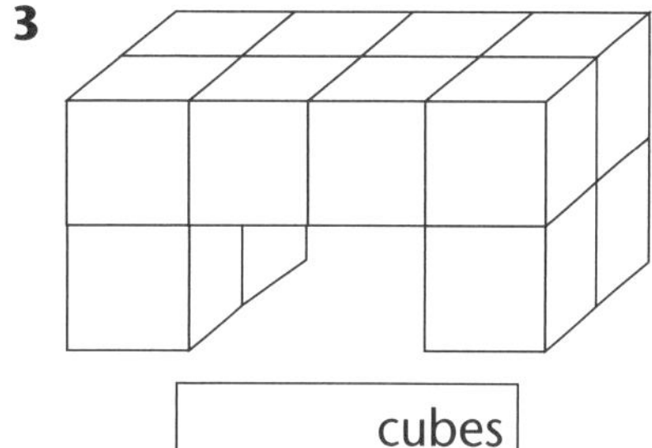

[ ] cubes

**4**

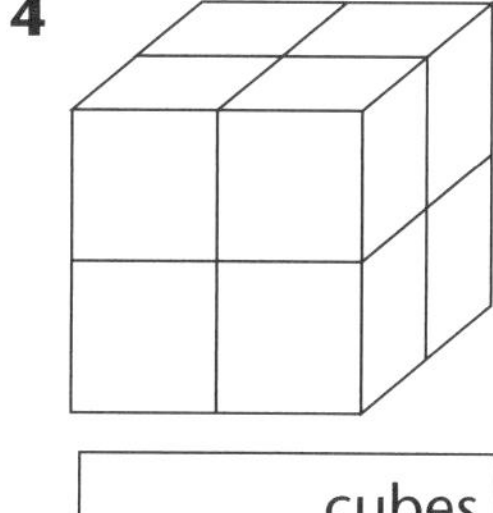

[ ] cubes

# Number and Algebra

## SET 1 Basic

1 6 × 8

2 19 – 10

3 21 + 4

4 40 ÷ 4

5 8 × 5

6 31 – 12

7 21 + 6 + 7

8 What is the sum of 19 and 6?

9 What is the product of 8 and 10?

10 3, 6, ☐, 12, ☐, ☐, 21

11 A quarter of 20

12 Ten less than 50

13 Write five in Roman numerals.

14 \$2.02 = ☐ cents

15 8694 = ☐ thousands + ☐ hundreds + ☐ tens + ☐ ones

16

How much do I have left from \$50 if I spent \$14.95?

\$ ☐

## SET 2 Solving word problems

1 Jack bought 6 shirts at \$30 each. How much did he spend?

2 275 children sat on chairs in the hall and 225 stood up. How many children were in the hall?

3 Tina saved \$268 and Fred saved \$86. How much did they have altogether?

4 Ken had seven \$10 notes, three \$5 notes and \$17 worth of coins. How much did he have altogether?

5 Lauren bought 8 boxes of chicken at \$7 each plus 4 drinks at \$3 each. How much did she spend?

6 Jack had \$87 and Jim had \$33 more than Jack. If Lisa had \$70, how much money did they have altogether?

# Number and Algebra Spreadsheets

Cahlie bought these items using her EFTPOS card.

Skirt
Cheer girl \$17

T-shirt
Bella's \$11

Shorts
Style Mart \$20

Shoes
Walkaway \$35

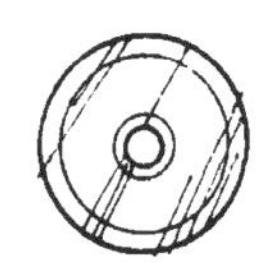

CD
Musica \$25

Handbag
Best Bags \$45

Complete the balance column on Cahlie's bank statement.

| | Date | Purchase | Debit | Balance |
|---|---|---|---|---|
| 1 | 1/10 | Opening | | \$198.00 |
| 2 | 2/10 | Skirt | \$17.00 | |
| 3 | 3/10 | T-shirt | \$11.00 | |
| 4 | 4/10 | Shorts | \$20.00 | |
| 5 | 5/10 | Shoes | \$35.00 | |
| 6 | 6/10 | CD | \$25.00 | |
| 7 | 7/10 | Handbag | \$45.00 | |

## Number and Algebra

### SET 3 Adding decimals

Shade the largest decimal.

| | | | |
|---|---|---|---|
| 1 | 0.5 | 0.7 | 0.2 |
| 2 | 0.6 | 0.9 | 0.3 |
| 3 | 0.34 | 0.58 | 0.41 |
| 4 | 2.30 | 2.75 | 2.12 |

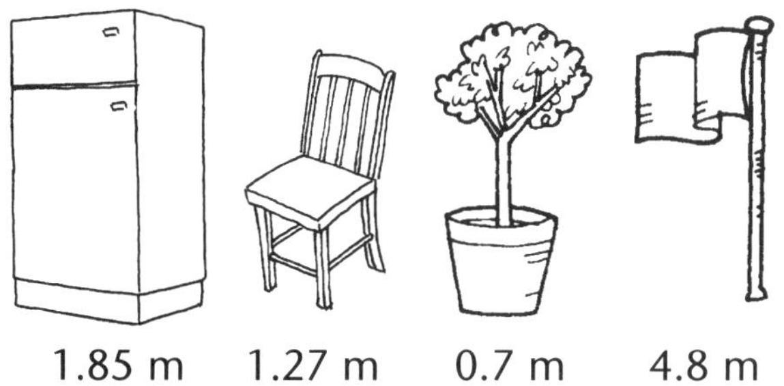

1.85 m 1.27 m 0.7 m 4.8 m

Solve the problems.

5 What is the total height of the refrigerator with a chair on top? 

6 What is the difference in height between a flagpole and a pot plant? 

### SET 4 Extension

1 3127, 3147, 3167, ☐

2 Grams in 4 kilograms

3 247 cm = ☐ m and ☐ cm

4 What number represents thousands in 4306?

5 What is the perimeter of a pentagon with 7 cm sides?

6 How many faces has a rectangular prism?

7 If 4 items cost \$3, how many would 8 cost?

8 Share \$28 among 7 people.

9 $42 \div 6 \times 2$

10 What is the change from \$5, if I spent \$1.85?

11 If 3 pears cost 60c, how much would 9 cost?

12 Write half-past five in digital time.

13 Is 40°C a hot day?

14 $\frac{1}{4}$ of 140

15 Write 497 in words.

16 Calculate the number of minutes between 1:55 am and 2:20 am

17 Round off 3743 to the nearest 100.

## Measurement Kilograms and grams

Label the mass shown on the scales.

1 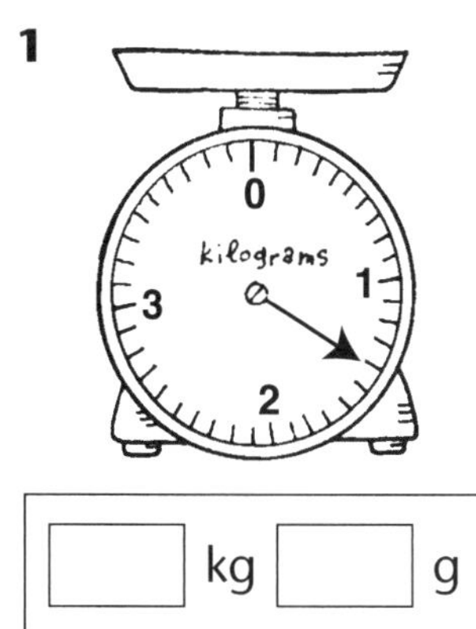


☐ kg ☐ g

2 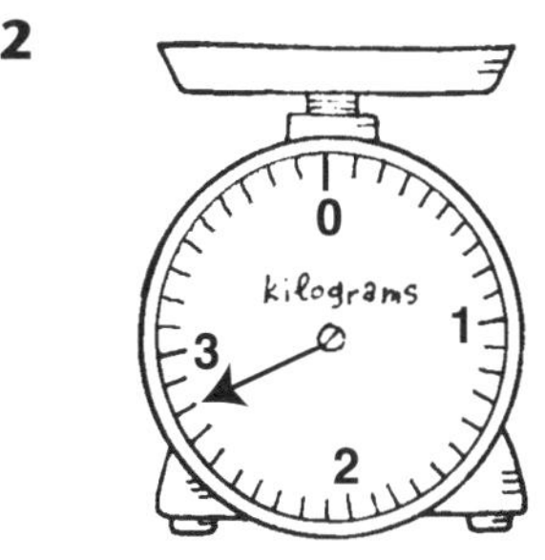


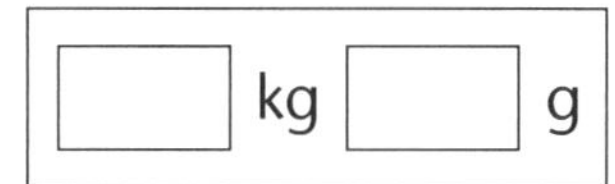

☐ kg ☐ g

3 

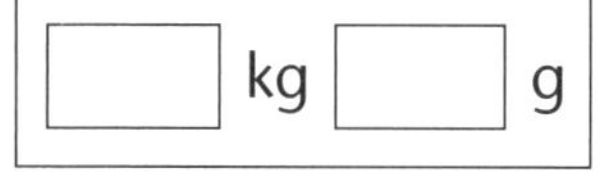

☐ kg ☐ g

4 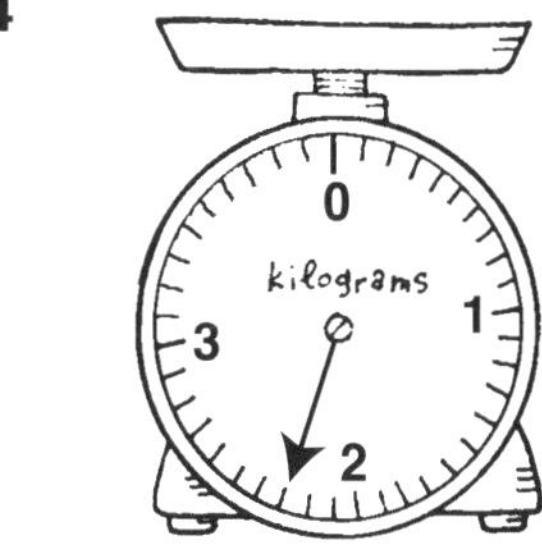


☐ kg ☐ g

UNIT 24

# Number and Algebra

## SET 1 Basic

**1** 6 × 4

**2** 20 – 12

**3** 20 + 18

**4** 50 ÷ 10

**5** 3 × 5

**6** 42 – 10

**7** 42 + 4 + 16

**8** What is the sum of 25 and 11?

**9** What is the product of 8 and 4?

**10** 90, 80, ☐, ☐, ☐, 40

**11** A third of 18

**12** Ten less than 48

**13** Write 12 in Roman numerals.

**14** 3076 = ☐ thousands + ☐ hundreds + ☐ tens + ☐ ones

**15**

What are the missing numbers? 1, 2, 4, ☐, 11, ☐, 22

## SET 2 Add and subtract

Add the numbers.

**1**
```
  2 3 4 5
+ 3 1 2 4
```

**2**
```
  4 2 3 7
+ 4 3 8 6
```

**3**
```
  3 5 7 6
+ 3 6 6 6
```

**4**
```
  8 5 7 9
+   6 3 4
```

Subtract these numbers.

**5**
```
  8 7 8 0
–   9 3 2
```

**6**
```
  7 6 5 4
– 2 8 8 3
```

**Mathematical Reasoning**

| | |
|---|---|
| Auckland – Brisbane | 2293 km |
| Brisbane – Sydney | 750 km |
| Sydney – Perth | 3274 km |

**7** What is the distance covered by a plane travelling from Auckland to Sydney via Brisbane? ☐

**8** What distance is covered by a plane travelling from Brisbane to Perth via Sydney? ☐

# Space Grid reference points

Draw the snooker balls on the table.

| | |
|---|---|
| Red balls | B5, B1, C3 |
| Black | H2 |
| Pink | B3 |
| Blue | E2 |
| Yellow | D4 |
| Green | H5 |
| Brown | H4 |
| White | I5 |

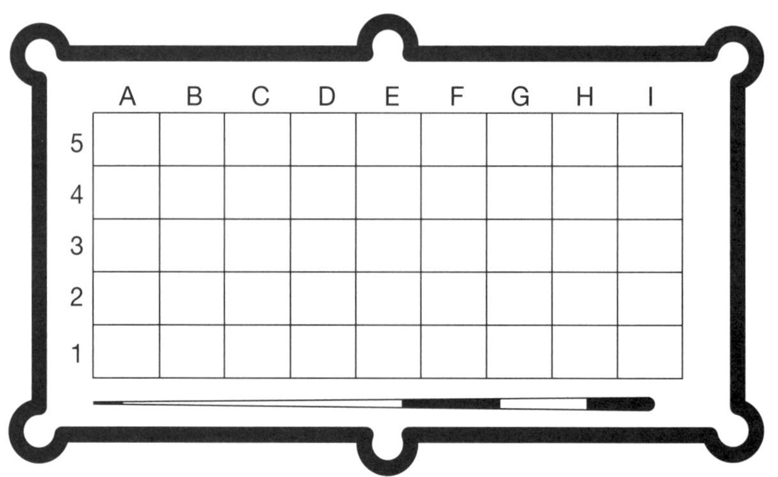

# Number and Algebra

## SET 3 Partitioning/multiplication

Complete the multiplications.

1. $\begin{array}{r} 35 \\ \times\ 3 \\ \hline \end{array}$
2. $\begin{array}{r} 43 \\ \times\ 4 \\ \hline \end{array}$
3. $\begin{array}{r} 29 \\ \times\ 5 \\ \hline \end{array}$
4. $\begin{array}{r} 36 \\ \times\ 5 \\ \hline \end{array}$
5. $\begin{array}{r} 27 \\ \times\ 6 \\ \hline \end{array}$
6. $\begin{array}{r} 46 \\ \times\ 8 \\ \hline \end{array}$

| | | |
|---|---|---|
| 7 | Zena saved $43 per week for 5 weeks. How much did she save altogether? | |
| 8 | Mr Caroulis planted 36 trees in each paddock. How many trees did he plant if there were 6 paddocks? | |

## SET 4 Extension

1. How many 75 cm lengths in 3 m?
2. What is the difference between 187 and 45?
3. 7 decades = ☐ years
4. 3 × 4 + 7
5. How many corners has a cube?
6. Does a decagon have 10 sides?
7. 7 pears at 85c each
8. Write all factors for 18.
9. $\frac{1}{4}$ of $96
10. 1375 mL = ☐ L + ☐ mL
11. Is 12 a multiple of 6?
12. What is the value of 0 in 7603?
13. What is the smallest number of coins needed to make $4.65?

**Mathematical Reasoning**

14. If a car travels at 50 km/h for 4 hours, how far would it travel?
15. How far will a train travel in 4 hours at 100 km/h?

# Space Rotational symmetry

Colour in the shapes that have rotational symmetry. List the order of rotational symmetry in the boxes.

1 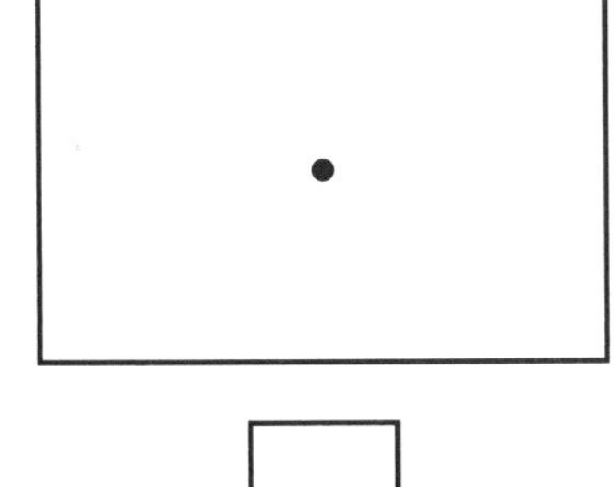

2 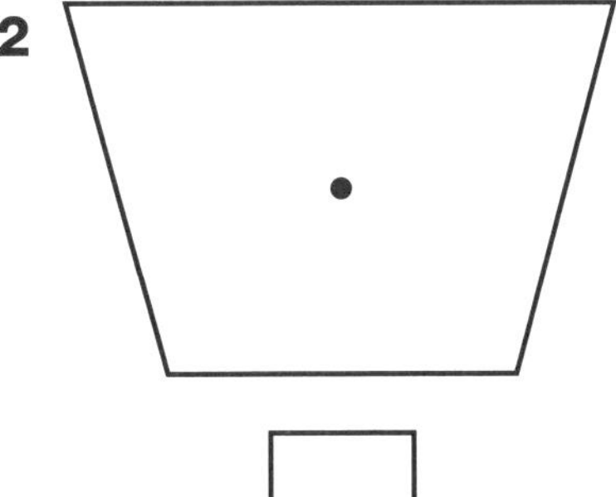

3 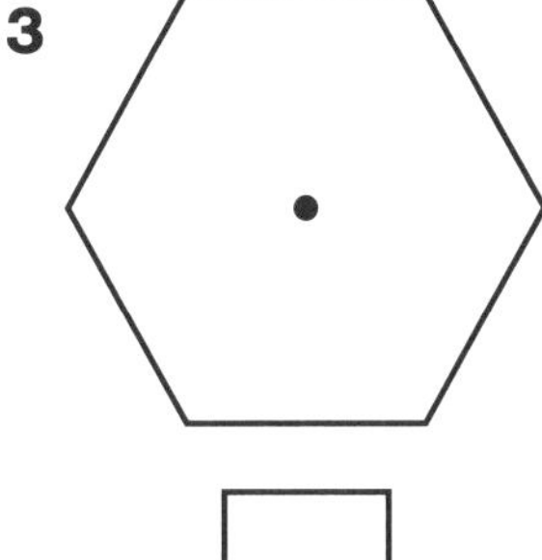

UNIT 25

# Number and Algebra

## SET 1 Basic

1 $7 \times 7$

2 $17 - 6$

3 $17 + 6$

4 $14 - 7$

5 $8 \times 7$

6 $6 + 6 + 6 + 6 - 7$

7 $9 \times 7$

8 What is the product of 7 and 10?

9 What is the difference between 49 and 20?

10 What is the product of 7 and 4?

11 How many 5c coins in 65c?

12 A quarter of 16

13 $2.59 = ☐ cents

14 320 = ☐ thousands + ☐ hundreds + ☐ tens + ☐ ones

15

What odd number is between 40 and 46 and greater than 43? ☐

## SET 2 Partitioning/multiplication

1 $23 \times 4$

2 $24 \times 5$

3 $35 \times 5$

4 $23 \times 6$

5 $16 \times 6$

6 $24 \times 6$

7 $35 \times 2$

8 $38 \times 3$

9 $28 \times 4$

10 Billy bought 7 CDs that cost $25 each. How much did he spend?

11 Lisa bought 5 skirts for $37 each. How much did she spend?

# Statistics and Probability Surveys

Mrs Sampson's class surveyed and graphed the points scored by 6 teams.

1 Which teams are leading the competition?

2 Which team is coming last?

3 Which team has 200 points?

4 How many more points does GP need to join the leaders?

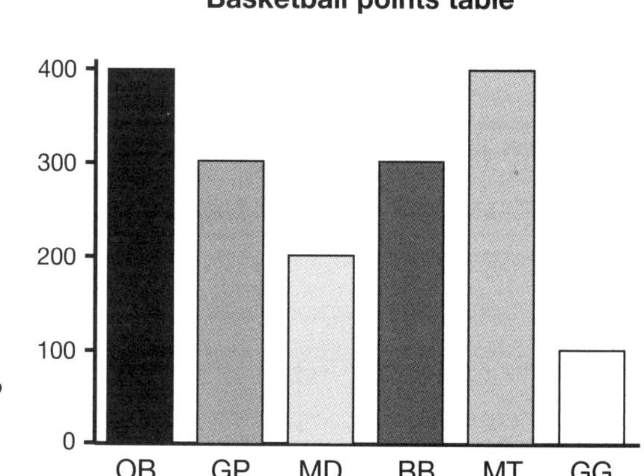

# Number and Algebra

## SET 3 Place value to tens of thousands

| tens hundreds thousands ones tens of thousands |
|---|

Use the words in the box to write the place value of the bold numbers.

| | | |
|---|---|---|
| 1 | 35**7**72 | hundreds |
| 2 | 374**5**6 | |
| 3 | 3567**8** | |
| 4 | 5**6**434 | |
| 5 | 98**5**53 | |
| 6 | **2**3657 | |
| 7 | 4**4**538 | |

Write these numbers.

**8** Twenty-three thousand, two hundred and ninety-six ____________

**9** Thirty-five thousand, four hundred and thirteen ____________

**10** Fifty-four thousand and eighty-six ____________

## SET 4 Extension

**1** Estimate an answer to 137 + 203.

**2** $3\frac{1}{4}$ kg = ☐ g

**3** How many 250 g bags in 2 kg?

**4** Half of a kilometre = ☐ m

**5** Share 40 among 6.

**6** How many quarters in $1\frac{3}{4}$?

**7** How many minutes in half an hour?

**8** $5^2 + 3$

**9** $5 \times 8 + 4$

**10** 4302 + 583

**11** 30 + 5 + 108

**12** What is the product of 9 and 6?

**13** Arrange 5, 6, 4, 9 to make the largest number possible.

**14** This model was built with 7 cubes. If Lea numbered every square on the model's surfaces, how many numbers did she use?

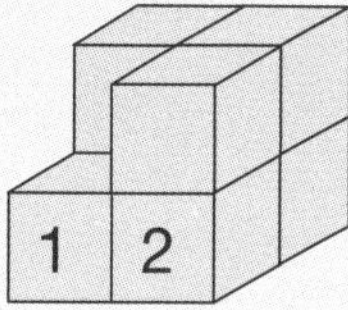


Mathematical Reasoning

# Measurement Litres and millilitres

Record these capacities in litres using decimal notation.

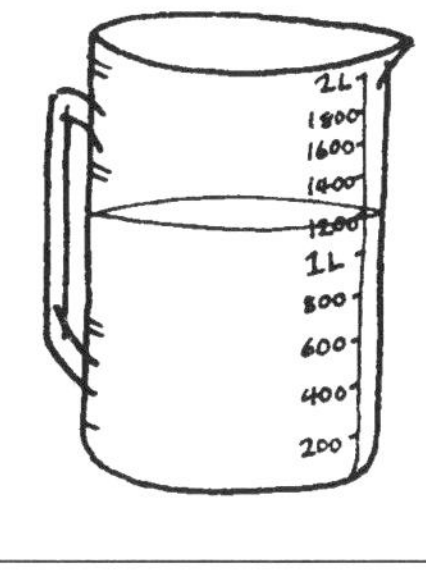


**1** ☐ . ☐ L

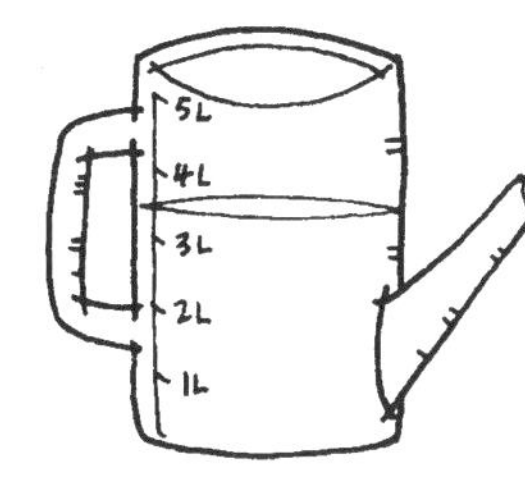


**2** ☐ . ☐ L

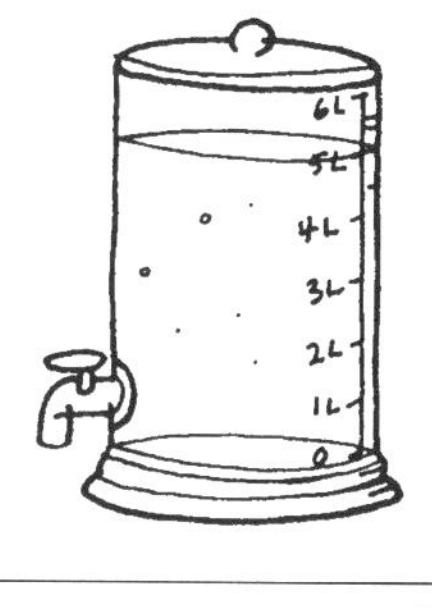


**3** ☐ . ☐ L

UNIT 26

# Number and Algebra

## SET 1 Basic

1 7 × 2

2 30 – 15

3 36 + 11

4 60 ÷ 6

5 8 × 8

6 58 – 10

7 56 + 14

8 What is the sum of 37 and 13?

9 8, 16, ☐, 32

10 What is a third of 24?

11 Ten less than 138

12 A quarter of 80

13 $3.79 = ☐ cents

14 Write 176 in words.

☐

15

## SET 2 Multiplication

$27 $34 $53

Use any strategy you wish to solve the multiplications.

| | |
|---|---|
| 1 How much would 3 CDs cost? | 2 How much would 4 books cost? |
| 3 How much would 5 shirts cost? | 4 How much would 6 CDs cost? |

5 Con bought 4 CDs and 2 shirts. How much money did he spend? 

6 Sally bought 6 shirts and a CD. How much did she spend?

## Space Angles

Colour the larger angle in each pair.

1

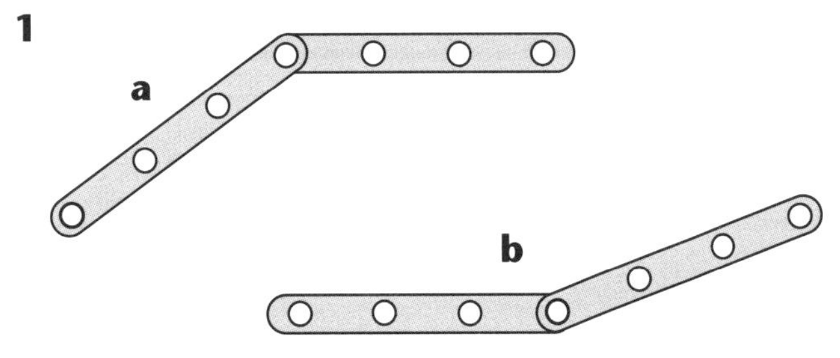


2

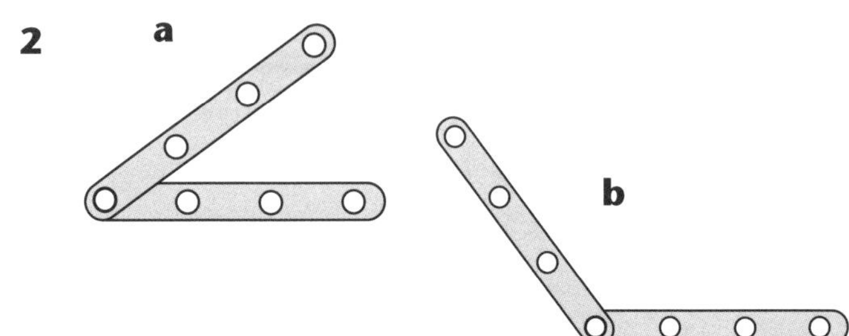

# Number and Algebra

## SET 3 Missing numbers

Complete the number sentences.

| | | | | | | | |
|---|---|---|---|---|---|---|---|
| 1 | 35 | + | | = | 50 | | |
| 2 | 60 | + | | = | 100 | | |
| 3 | 90 | – | | = | 60 | | |
| 4 | 40 | – | | = | 15 | | |
| 5 | 86 | – | | = | 60 | | |
| 6 | 6 | × | | = | 18 | | |
| 7 | 7 | × | | = | 28 | | |
| 8 | 6 | × | | = | 42 | | |
| 9 | 50 | ÷ | | = | 10 | | |
| 10 | 30 | ÷ | | = | 5 | | |
| 11 | 7 | × | 8 | + | | = | 60 |
| 12 | 6 | × | 4 | + | | = | 40 |
| 13 | 5 | × | 8 | + | | = | 60 |
| 14 | 40 | ÷ | 5 | + | | = | 60 |

## SET 4 Extension

1. \$546 – \$12
2. $4\frac{1}{4}$ kg = ☐ g
3. List the factors of 15.
4. $\frac{1}{10}$ of a kilogram = ☐ g
5. Share 50 among 6 people.
6. How many tenths in a whole?
7. How many minutes in $\frac{3}{4}$ of an hour?
8. Round 1969 to the nearest 100.
9. $6^2 + 3$
10. List the factors of 20.
11. If 9 pencils cost 81c, how much would 20 pencils cost?
12. What is the product of 9 and 4?
13. Arrange 6, 1, 7, 2 to make the smallest number possible.
14. If 5 rulers cost 95c, how much would 20 cost?

**Mathematical Reasoning**

15. Lee is 5 years older than Jack, who is 15 years younger than Nick, who is 2 years older than Con.

    How old is Lee if Con is 45? ________

# Measurement am and pm time

Write a digital label for each clock using am and pm notation.

**1** morning

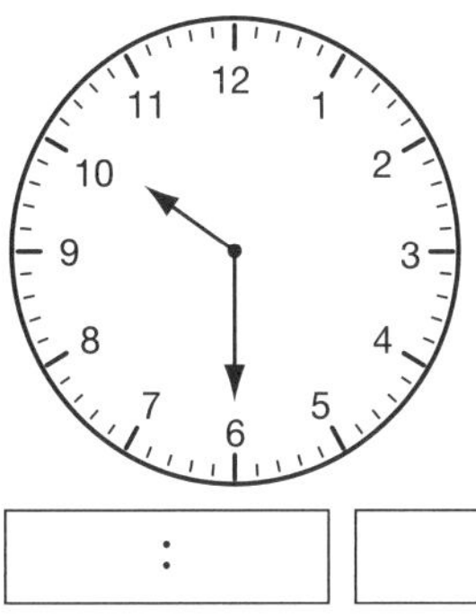

☐ : ☐ ☐

**2** morning

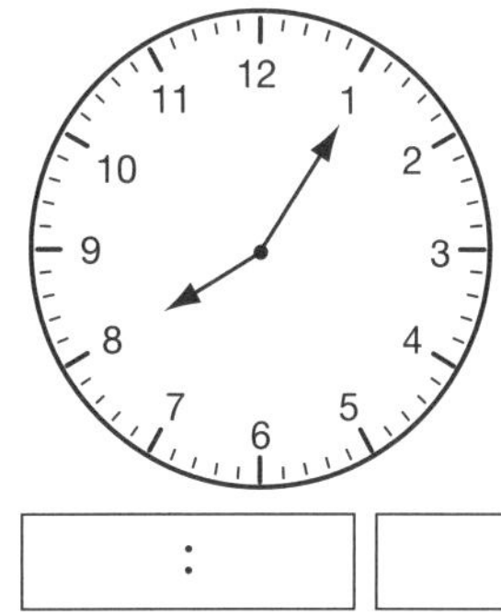

☐ : ☐ ☐

**3** afternoon

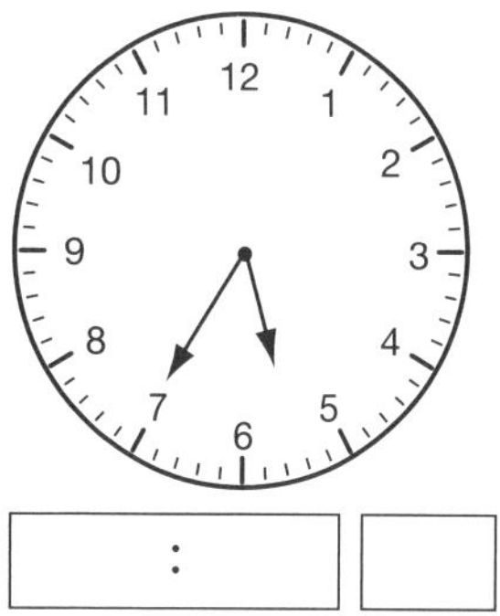

☐ : ☐ ☐

**4** evening

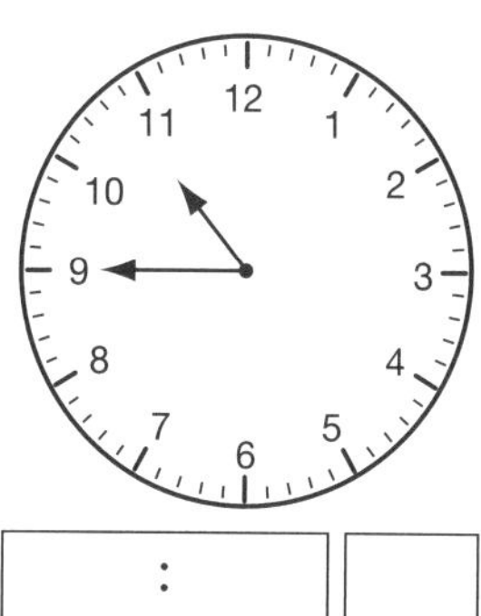

☐ : ☐ ☐

# Number and Algebra

## SET 1 Basic

1 50 – 25

2 10 – 8

3 28 – 21

4 24 + 9

5 18 + 6

6 9 × 3

7 9 × 5

8 9 × 4

9 What is the product of 7 and 9?

10 What is the sum of 17 and 13?

11 A third of 30

12 309c = $ ☐

13 What is the difference between 40 and 25?

14 27 = ☐ thousands + ☐ hundreds + ☐ tens + ☐ ones.

15

How old am I if I'm 16 years younger than Aunt Kylie, who is 25? ☐

## SET 2 Number patterns

Follow the rules to complete the number patterns.

1

| Multply by 4 | | | | | | | |
|---|---|---|---|---|---|---|---|
| 2 | 4 | 6 | 8 | 10 | 12 | 14 | 16 |
| | | | | | | | |

2

| Adding 15 | | | | | | | |
|---|---|---|---|---|---|---|---|
| 1 | 2 | 3 | 4 | 5 | 6 | 7 | 8 |
| | | | | | | | |

3

| Multiplying by 4 | | | | | | | |
|---|---|---|---|---|---|---|---|
| 1 | 2 | 3 | 4 | 5 | 6 | 7 | 8 |
| | | | | | | | |

4

| Multiply by 6 | | | | | | | |
|---|---|---|---|---|---|---|---|
| 3 | 4 | 5 | 6 | 7 | 8 | 9 | 10 |
| | | | | | | | |

Find the rule and complete the pattern.

5

| Rule: | | | | | | | |
|---|---|---|---|---|---|---|---|
| 1 | 2 | 3 | 4 | 5 | 6 | 7 | 8 |
| 5 | 6 | 7 | | | | | |

6

| Rule: | | | | | | | |
|---|---|---|---|---|---|---|---|
| 1 | 2 | 3 | 4 | 5 | 6 | 7 | 8 |
| 3 | 6 | 9 | | | | | |

# Space Quadrilaterals

1 Colour the parallelograms.

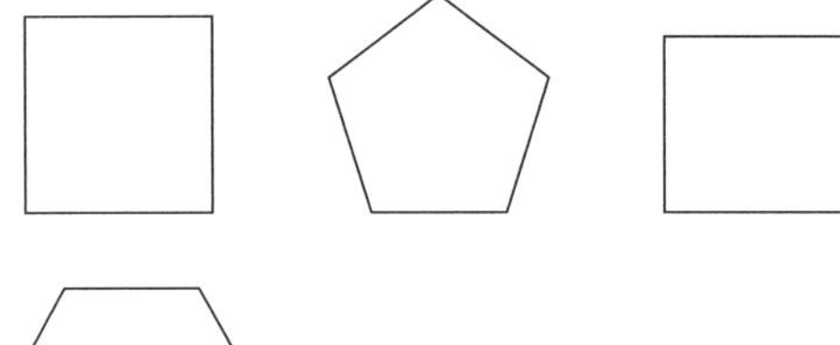

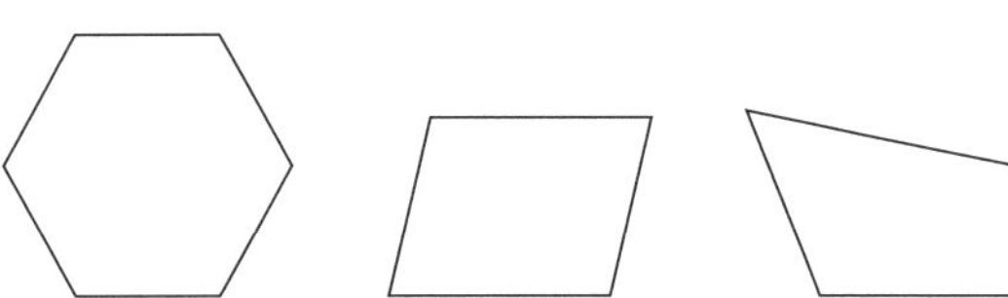

2 Colour the shape that matches the description.

I have 4 sides of equal length. I have 2 angles smaller than a right angle and 2 angles greater than a right angle.

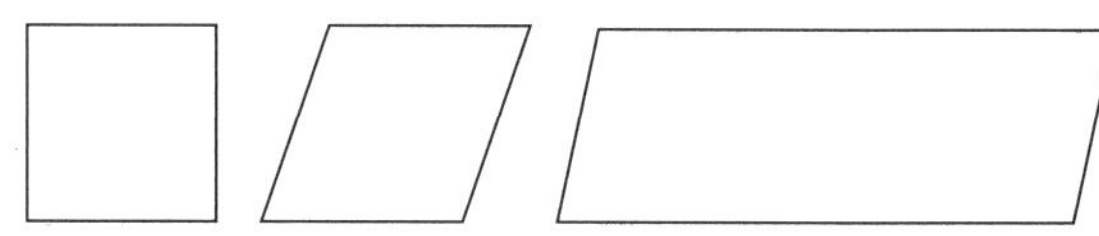

# Number and Algebra

## SET 3 Partitioning numbers

Expand the numbers.

**1** 35 674

30 000 + ___ + ___ + ___ + ___

**2** 56 978

___ + ___ + ___ + ___ + ___

**3** 68 926

___ + ___ + ___ + ___ + ___

**4** 74 854

___ + ___ + ___ + ___ + ___

**5** 69 276

___ + ___ + ___ + ___ + ___

**6** 65 273

**7** 74 541

**8** 85 962

## SET 4 Extension

**1** $(\frac{1}{2}$ of $24) \times 0$

**2** How many corners has a rectangular prism?

**3** $5^2 + 5$

**4** 3000 mL = ___ L

**5** Is 42 a multiple of 6?

**6** 21 hundredths = 0.___

**7** What is the value of 9 in 9327?

**8** How much are 3 pieces of cake at 35c each?

**9** How much is $2\frac{1}{2}$ kg of flour at $1.20 per kg?

**10** Write all the factors for 30.

**11** How many millilitres in $5\frac{1}{2}$ L?

**12** If 8 cost $96, how much would 5 cost? 

**13** A boy ran for 3 hours at an average speed of 12 km/h. How far did he run? 

**Mathematical Reasoning**

**14** Gillian is trying to solve 56 ÷ 8 on her calculator but the 8 button is missing.

Place numbers in the boxes so that her problem can be solved.

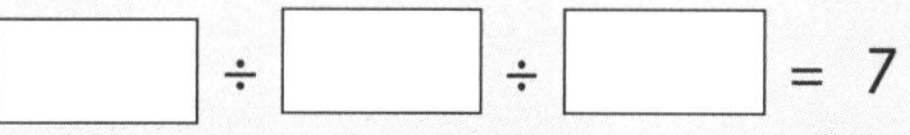

# Statistics and Probability Combinations

Jim's soccer team has a choice of colours for their new uniform. They can have red, yellow or blue tops combined with yellow or blue shorts. Colour all combinations of uniform they could have.

UNIT
28

## Number and Algebra

### SET 1 Basic

1 38 – 24

2 What is the product of 6 and 4?

3 20 ÷ 4

4 6, 12, ☐, 24, ☐

5 Ten less than 254

6 45 ÷ 5

7 What is the difference between 38 and 19?

8 A quarter of 20

9 $3.65 = ☐ c

10 42 ÷ 6

11 A third of 21

12 What is the sum of 8, 9 and 10?

13 Eleven less than 44

14 659c = $ ☐.☐☐

15 Write four thousand, three hundred and twenty-one as a numeral.

16 

Chris saved $6 a week for 8 weeks. How much did he save altogether?

$ ☐

### SET 2 Rounding to the nearest 5c

Round these amounts to the nearest 5c.

1 $0.41

2 $0.52

3 $0.38

4 $0.49

5 $0.63

6 $0.78

7 $1.52

8 $2.73

9 $3.56

10 $4.58

Solve the problems and round the answers to the nearest 5c.

11 Jason bought a stamp for 55c and a pencil for 44c. How much did he spend?

12 Sarah bought a book for $1.38 and a pen for $0.56. How much did she spend? 

## Statistics and Probability Investigating likelihood

Taylor has a bag of 11 marbles coloured red, blue, pink and green.

1 Which marble is more likely to be drawn out of the bag? __________

2 Which marbles are least likely to be drawn out of the bag? ________

3 If 3 red marbles were drawn out of the bag and thrown away, what marble would then be most likely to be drawn out of the bag? _____

4 Is it more likely that a pink marble would be pulled out than a blue one? ______

# Number and Algebra

## SET 3 Decimal fractions

Write each measurement as a decimal.

| | | |
|---|---|---|
| 1 | 1 m and 26 cm | . m |
| 2 | 1 m and 75 cm | . m |
| 3 | 2 m and 37 cm | . m |
| 4 | 1 m and 58 cm | . m |
| 5 | 3 m and 99 cm | . m |
| 6 | 5 m and 65 cm | . m |
| 7 | 6 m and 7 cm | . m |

Order these decimals from smallest to largest.

| | | | |
|---|---|---|---|
| 8 | 1.36 m | 2.07 m | 1.07 m |
| | | | |
| 9 | 3.23 m | 1.32 m | 2.31 m |
| | | | |
| 10 | 2.54 m | 5.42 m | 5.24 m |
| | | | |

## SET 4 Extension

1 \$6.25 + \$4.50

2 6 kg at \$4.20 per kg

3 How many tens in 568?

4 Round 1944 to the nearest 10.

5 54 ÷ 6 + 8

6 1, 4, 9, ☐, 25, ☐, 49

7 What is the change from \$5 if I spent \$2.45?

8 How many thousands in 85 901?

9 Share 48 among 5.

10 What is the value of 6 in 16 250?

11 Three watches at \$99 each

12 8 decades = ☐ years

13 Round off to the nearest 10 to find an estimate for 687 – 253.

14 How many 60 cm lengths can be cut from 3 m?

15 Mr Klim spent \$36 on a book and a calculator. How much did the book cost if the calculator cost twice as much?

Book \$ ☐

Mathematical Reasoning

# Number and Algebra Multiplication and division number patterns

Follow the rules to complete the patterns then use backtracking to check them.

1

| × 3 | 2 | 4 | 6 | 8 | 10 |
|---|---|---|---|---|---|
| | | | | | |

2

| ÷ 5 | 15 | 25 | 35 | 45 | 55 |
|---|---|---|---|---|---|
| | | | | | |

3

| ÷ 4 | 40 | 36 | 32 | 28 | 24 |
|---|---|---|---|---|---|
| | | | | | |

4

| × 8 | 1 | 3 | 5 | 7 | 9 |
|---|---|---|---|---|---|
| | | | | | |

# Number and Algebra

## SET 1 Basic

1 4 × 10c

2 5c + 15c

3 3 × $2

4 $20 ÷ 2

5 $18 + $12

6 $4.50 – $2.00

7 $5 × 6

8 $21 ÷ 3

9 What is the product of 6 and 11?

10 What is the sum of 15 and 14?

11 What is the difference between 35 and 42?

12 One quarter of $36

13 Share $24 among 3 people.

14 Write $9.56 in words. ____________________

____________________________________

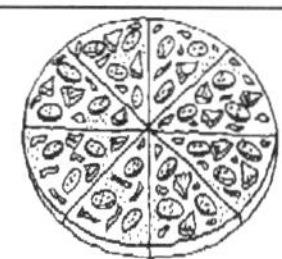

15 We ordered two pizzas. How much pizza is left if we ate one half of each pizza?

## SET 2 Multiplying tens

1 40 × 9

2 80 × 5

3 50 × 2

4 30 × 8

5 20 × 7

6 50 × 4

7 60 × 6

8 70 × 7

9 70 × 5

10 90 × 9

Supply the missing number.

11 50 × ☐ = 300

12 60 × ☐ = 540

13 ☐ × 3 = 240

14 ☐ × 4 = 280

How much money is in each bundle?

15 

$

16 

$

17 

$

# Statistics and Probability Picture graphs

Construct a picture graph from the given data. It has been started for you.

**Children's ages**

| Age | Number |
|---|---|
| 8 | ЖЖ ЖЖ II |
| 9 | ЖЖ ЖЖ ЖЖ |
| 10 | ЖЖ ЖЖ II |
| 11 | ЖЖ IIII |

**Children's ages**

| 8 | 9 | 10 | 11 |
|---|---|---|---|

**Ages**

**Key**
☺ = 3 children

# Number and Algebra

## SET 3 Mixed addition

**1**

```
  3 5 7
      7
  2 0 6
+ 3 0 5
-------
```

**2**

```
  3 . 2 1
  2 . 5 4
+ 3 . 0 1
---------
```

**3**

```
  $ 3 5 . 2 4
+     1 2 . 6 9
---------------
```

**4**

```
  $ 5 6 . 3 9
+       7 . 0 8
---------------
```

**5**

```
  2 3
  3 4
  3 5
  1 6
+   7
-----
```

**6**

```
  3 5
  4 2
    2
    3
+ 7 5
-----
```

**7** Sarah bought a $26.95 pen and a $11.55 book. How much change did she receive from $50.00?

## SET 4 Extension

**1** $3 \times 0 \times 5$

**2** 37 hundredths = 0.☐

**3** $3\frac{1}{2}$ m = ☐ cm

**4** 6 pencils at 45c each

**5** How many faces has a square pyramid?

**6** Write all the factors for 21.

**7** ☐ $\times 6 + 1 = 37$

**8** How much is 3 kg of sugar at $4.30 per kg?

**9** Is 42 a multiple of 7?

**10** What fraction of 12 is 6?

**11** What is the value of 7 in 3075?

**12** Write 3 tenths as a decimal.

**13** 1 tenth of 100

**14** How many 10c coins in $12.50?

**15** Perimeter of an octagon with 13 cm sides

**16** Complete the magic square.

| | | |
|---|---|---|
| | 8 | 18 |
| | | 20 |
| | | 10 |

**17** If 4 kg of meat cost $20.00, how much is 7 kg?

# Measurement The square metre

Tom was retiling his kitchen/family room. He drew square metres on the floor to calculate how many tiles to order.

**1** How many square metres of tiles does he need to cover his kitchen benches?

☐ square metres

**2** What is the area of the whole kitchen/family room?

☐ square metres

**Kitchen/family room**

benches

Kitchen

Fridge

UNIT 30

## Number and Algebra

### SET 1 Basic

1 $5 \times 11c$

2 $6c + 6c + 5c$

3 \$2 × 3

4 \$18 ÷ 6

5 \$72 + \$19

6 \$4.50 – \$3

7 \$15 × 2

8 \$21 ÷ 7

9 What is the product of 8 and 3?

10 What is the sum of 13, 18 and 19?

11 What is the difference between 49 and 100?

12 What is one fifth of \$10?

13 Share \$72 among 9 people.

14 Write \$16.58 in words.

___________________________________

___________________________________

### SET 2 Division

Solve the divisions.

1 $3\overline{)24}$ 2 $4\overline{)32}$ 3 $5\overline{)45}$

4 $3\overline{)48}$ 5 $4\overline{)28}$ 6 $6\overline{)72}$

7 $5\overline{)65}$ 8 $7\overline{)91}$ 9 $8\overline{)96}$

Use the halve and halve again strategy to solve these.

10 20 ÷ 4

11 44 ÷ 4

12 60 ÷ 4

13 48 ÷ 4

14 80 ÷ 4

15 68 ÷ 4

16 72 ÷ 4

17 88 ÷ 4

Use the halve, halve and halve again strategy to solve these.

18 24 ÷ 8

19 32 ÷ 8

20 48 ÷ 8

21 64 ÷ 8

22 80 ÷ 8

23 96 ÷ 8

24 104 ÷ 8

25 120 ÷ 8

## Space Grouping two-dimensional shapes

1 Sketch 2 different types of quadrilaterals (4-sided shapes).

2 Sketch 2 different types of triangles.

# Number and Algebra

## SET 3 4-digit subtraction

Answer the subtractions.

**1**

```
  9 6 7 4
- 3 2 5 6
_________
```

**2**

```
  2 5 7 6
-   5 4 8
_________
```

**3**

```
  6 5 3 2
-   2 0 5
_________
```

**4**

```
  7 4 5 3
- 5 0 7 6
_________
```

**5**

```
  6 2 0 4
- 1 6 6 5
_________
```

**6**

```
  3 8 0 0
- 1 6 7 8
_________
```

**7** Jack saved $8500 towards a new car. If he only spent $7326, how much does he have left?

**8** The theatre holds 4217 people. If 1406 people are already seated, how many seats are still empty?

## SET 4 Extension

**1** $92 – $13

**2** 4 L = ☐ mL

**3** List the factors of 36.

**4** Millilitres in quarter of a litre

**5** Share 50c among 10 people.

**6** How many hundredths in a whole?

**7** 6:15 am + 20 mins

**8** $6^2 - 5$

**9** $6 \times 9 - 4$

**10** List the factors of 24.

**11** How much is $3\frac{1}{2}$ kg of meat at $1.20 kg?

**12** What is the product of 8 and 6?

**Mathematical Reasoning**

**13** Use rounding strategies to estimate who trains for the longer time.

Sally trains for 55 minutes 5 days a week. Rod trains for 39 minutes 7 days a week.

Use a calculator to check your estimates.

# Measurement Duration/timetables

| CHANNEL 13 | |
|---|---|
| 12:00 | Breakers |
| 12:30 | Movie–Taboo |
| 2:00 | Dr Dill |
| 2:30 | World Sport |
| 3:00 | Joe's Cooking |
| 4:00 | Cartoon Corner |
| 5:00 | Totally Mild |
| 5:30 | Price is Right |
| 6:00 | News |

Answer the questions about Channel 13's afternoon programs.

**1** At what time does the movie start? ______

**2** How long does the movie run for? ______

**3** What time does Joe's Cooking start? ______

**4** What is on at 5:30? ______

**5** What is on at 2:00? ______

**6** How long does Cartoon Corner run for? ______

# Number and Algebra

## SET 1 Basic

**1** 16 – 5

**2** 6 × 8

**3** 6 + 8

**4** 7 × 8

**5** 8 × 4

**6** 10 + 10 + 8 – 18

**7** What is the sum of 18 and 12?

**8** What is the product of 9 and 9?

**9** How many threes in 24?

**10** Half of $26

**11** $13.50 = ☐ c

**12** 2706 = ☐ thousands + ☐ hundreds + ☐ tens + ☐ ones

**13** What is the difference between 36 and 14?

**14** 3 × 7 + 9

**15**

I have 21 points. How many more do I need to equal the record, which is 57? ☐

## SET 2 Division with remainders

Answer the divisions.

**1** 7 ÷ 2 = ☐ r ☐

**2** 16 ÷ 3 = ☐ r ☐

**3** 18 ÷ 4 = ☐ r ☐

**4** 23 ÷ 5 = ☐ r ☐

**5** 33 ÷ 6 = ☐ r ☐

**6** 50 ÷ 7 = ☐ r ☐

**7** 33 ÷ 8 = ☐ r ☐

**8** 30 ÷ 9 = ☐ r ☐

**9** Farmer Jackson shared 36 sheep into 6 paddocks. How many sheep were in each paddock?

**10** Jenny had 48 matches that she put into 6 boxes. How many matches did she put in each box?

**Mathematical Reasoning**

Write a division question for the answers.

**11** ☐ ÷ ☐ = 2 r 1

**12** ☐ ÷ ☐ = 3 r 2

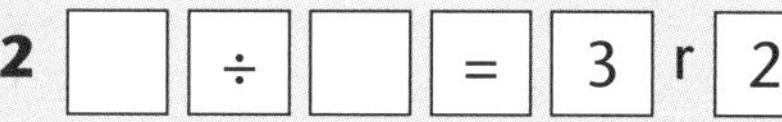

# Space Tessellations

Continue this tessellating pattern.

# Number and Algebra

## SET 3 Addition and subtraction problems

Answer the questions.

**1**
```
  3 5 7 9
+ 2 6 0 8
---------
```

**2**
```
  3 5 6 4
+ 3 8 0 9
---------
```

**3**
```
  6 5 9 8
+ 2 8 7 0
---------
```

**4**
```
  3 6 8 7
+   6 6 7
---------
```

**5**
```
  3 5 2 0
- 1 7 0 6
---------
```

**6**
```
  6 3 0 7
-   3 8 8
---------
```

**7** Lisa bought a desk for $1037, a bookcase for $835 and a chair for $340. How much did she spend?

**8** Lenny saved $1426 one year and $1820 the next. How much does he have left from his savings if he spent $2056 on a holiday?

## SET 4 Extension

**1** How many grams in $1\frac{1}{4}$ kg?

**2** Share 42 among 3 people.

**3** Five books at $41 each

**4** What is the value of 1 in 3015?

**5** 200 mL = 2 L True/False?

**6** Write the smallest number you can using 1, 8, 7, 3.

**7** How many tenths in 2 wholes?

**8** 37 hundredths = [0.      ]

**9** Write 1 whole and 3 tenths as a decimal.

**10** How many centimetres in 3.25 m?

**11** Is 128 a multiple of 7?

**12** 6 cakes cost $36, how much for 11?

**13** How much is $2\frac{1}{2}$ kg of chicken at $6 a kilogram?

**Mathematical Reasoning**

**14** Jessica bought two items. If the iPod cost $15 less than the stereo, how much change would she receive from $200?

# Statistics and Probability Representing data

**1** Ms Toms' class conducted a survey to find out which cars were driven by the teachers. Use the table below to construct a bar graph.

| Make | Ford | Toyota | Holden | Nissan | Mazda | Other |
|---|---|---|---|---|---|---|
| Number of cars | 𝍸 𝍸 | 𝍸 ||| | 𝍸 𝍸 | |||| | || | 𝍸 ||| |

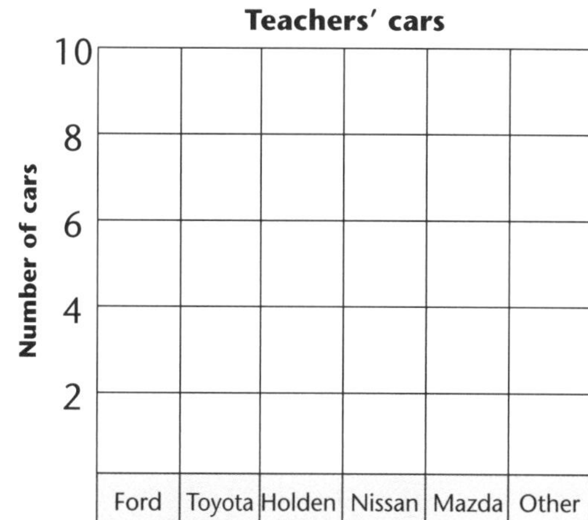


**2** What were the two most popular makes of cars driven by the teachers?

# Number and Algebra

## SET 1 Basic

1 $6 \times 5$

2 What is the product of 6 and 4?

3 3 mL + 10 mL

4 Triple 10c.

5 \$1.70 – 20c

6 228 cm = ☐ m ☐ cm

7 10 kg – 3 kg

8 Square 8.

9 21 m ÷ 3

10 What is the product of 9 and 9?

11 \$11 – \$10.50

12 42 m ÷ 6

13 20 mL + 25 mL

14 $21 \times 4$

15 

## SET 2 2-digit × 1-digit multiplication

Solve these multiplications.

1 $64 \times 2$

2 $86 \times 3$

3 $48 \times 3$

4 $35 \times 4$

5 $36 \times 5$

6 $47 \times 6$

7 Benny made 3 payments of \$28. How much did he spend? 

8 Jessica saved \$69 per month for 6 months. How much did she save? 

9 Jim's sister spilt ink on his homework. Write in the missing numbers to help Jim.

| | HUND | TENS | ONES |
|---|---|---|---|
| | | $\frac{2}{5}$ | [ink] |
| × | | | 6 |
| | [ink] | 2 | 4 |

Mathematical Reasoning

# Space Nets

Match the following objects to their nets.

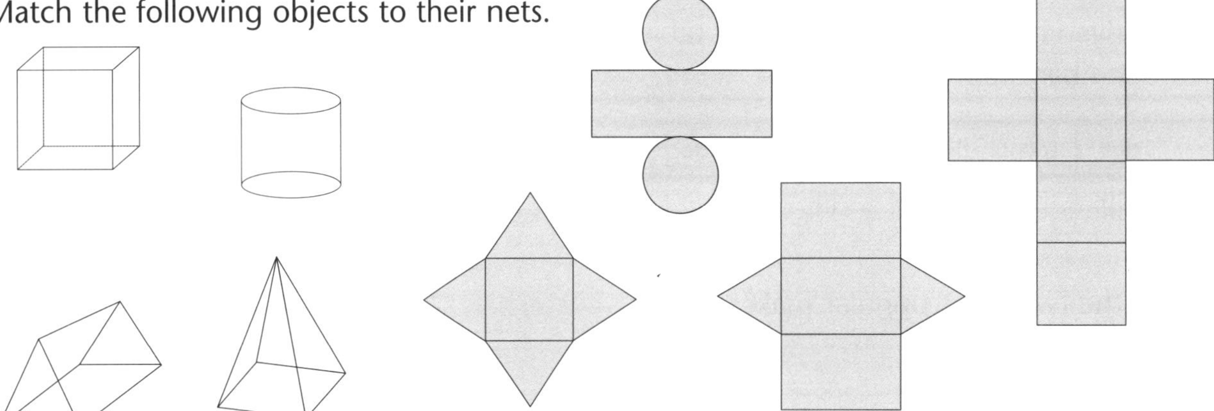

# Number and Algebra

## SET 3 Comparing fractions

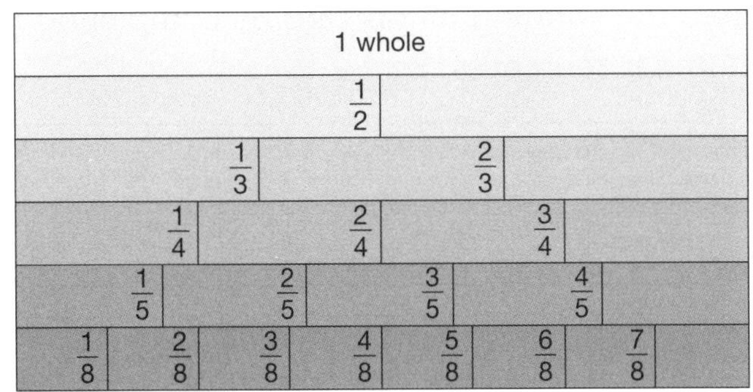


Use the fraction grid to answer true or false to the questions below.

$>$ greater than
$<$ less than

1 $\frac{1}{2} = \frac{4}{8}$ ______

2 $\frac{1}{2} = \frac{2}{4}$ ______

3 $\frac{1}{3} > \frac{1}{4}$ ______

4 $\frac{2}{3} > \frac{1}{2}$ ______

5 $\frac{3}{4} < \frac{1}{2}$ ______

6 $\frac{3}{4} < \frac{7}{8}$ ______

7 $\frac{6}{8} = \frac{2}{3}$ ______

8 $\frac{3}{8} > \frac{1}{4}$ ______

9 $\frac{5}{8} < \frac{3}{5}$ ______

## SET 4 Extension

1 $3 \times 7 + 0$

2 $3 \times 6 + 12 = 30$. True/False?

3 Round 4506 to the nearest 1000.

4 Centimetres in 4.27 m

5 What is the value of 8 in 11 874?

6 Write 7 tenths as a decimal.

7 Write 17 hundredths as a decimal.

8 If 9 apples cost $8.10, how much would 5 cost?

9 Is 48 a multiple of 9?

10 John's car averaged 100 km/h over $4\frac{1}{2}$ hours. How far did it travel?

11 Write all the factors for 35.

12 Write 7 hundredths as a decimal.

13 Complete the path.

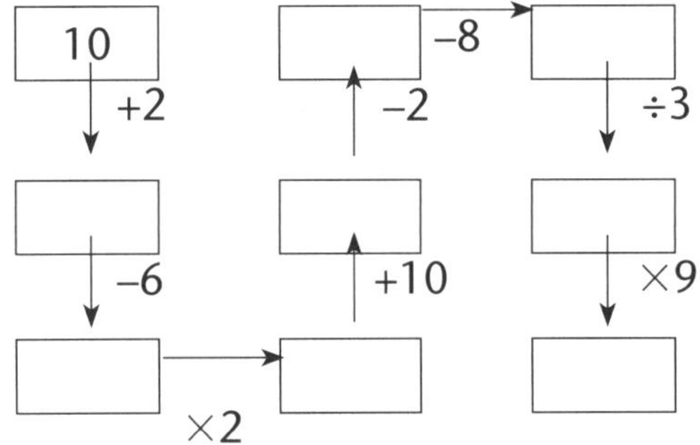


# Measurement Millimetres

Measure the length of these lines in millimetres.

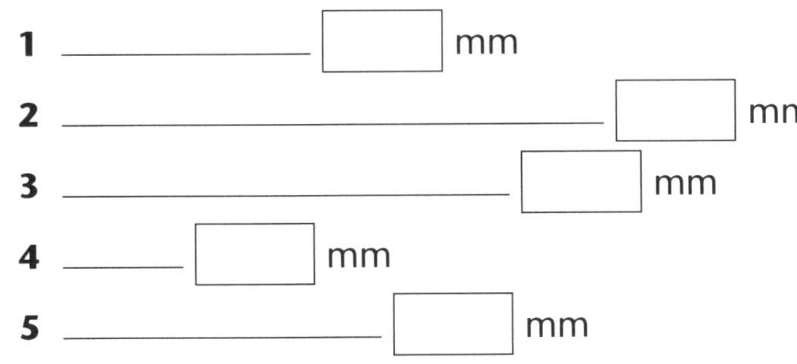


Convert each centimetre measurement into millimetres.

6 3 cm = ______ mm

7 8 cm = ______ mm

8 4 cm = ______ mm

9 $\frac{1}{2}$ cm = ______ mm

10 $5\frac{1}{2}$ cm = ______ mm

UNIT
33

# Number and Algebra

## SET 1 Basic

**1** 3 × 6

**2** 6 + 6 + 8

**3** 3 × 8

**4** 17 – 6

**5** 45 – 6

**6** 18 + 4

**7** 3 × 9

**8** 3 × 7

**9** What is the product of 7 and 5?

**10** How many days in August?

**11** What is the sum of 39 and 6?

**12** $31 – $7

**13** What is the difference between 37 and 20?

**14** 8756 = ☐ thousands + ☐ hundreds + ☐ tens + ☐ ones.

**15**

If 2 drinks cost $1.20, how much would 3 drinks cost?

## SET 2 Division

Find the winning Bingo card by completing the number sentences.

**a**

|  | 8 |  | 10 r 1 |  | 6 |  |
|---|---|---|---|---|---|---|
| 5 r 4 |  | 8 r 1 |  | 9 |  | 3 |

**b**

|  | 8 |  | 3 |  | 5 r 4 |  |
|---|---|---|---|---|---|---|
| 6 |  | 9 r 2 |  | 10 r 1 |  | 2 |

**c**

|  | 9 r 2 |  | 10 r 1 |  | 8 |  |
|---|---|---|---|---|---|---|
| 3 |  | 9 |  | 6 |  | 15 |

**1** 12 ÷ 4 = ☐

**2** 44 ÷ 8 = ☐

**3** 33 ÷ 3 = ☐

**4** 26 ÷ 2 = ☐

**5** 40 ÷ 5 = ☐

**6** 36 ÷ 6 = ☐

**7** 25 ÷ 3 = ☐

**8** 27 ÷ 3 = ☐

**9** 47 ÷ 5 = ☐

**10** 31 ÷ 3 = ☐

Solve these divisions.

**11** $4\overline{)90}$ **12** $6\overline{)92}$ **13** $6\overline{)79}$

**14** $2\overline{)77}$ **15** $4\overline{)65}$ **16** $3\overline{)52}$

# Space Mapping

**1** Colour the space C4 blue.

**2** Colour the space B5 red.

**3** Colour the space D3 green.

**4** Put an X in G7.

**5** Draw a circle in H1.

**6** Put the letter A in D8.

**7** Put the letter E in E5.

**8** Draw a triangle in A7.

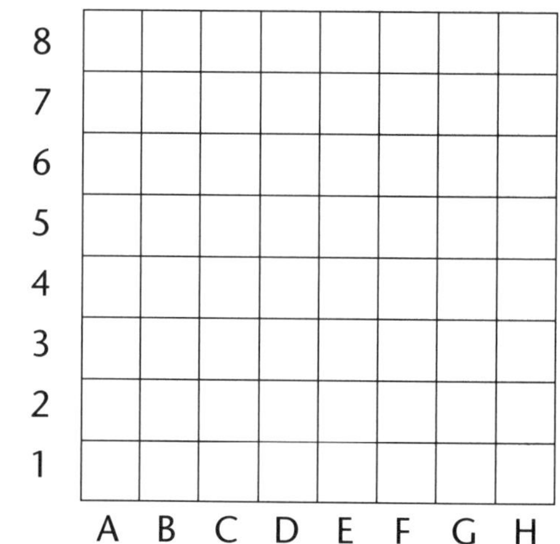

# Number and Algebra

## SET 3 Counting patterns with fractions

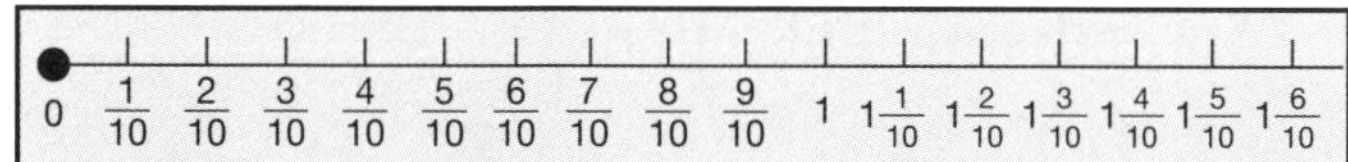


Use the number line to help you complete the patterns.

**1**

| $0$ | $\frac{1}{10}$ | $\frac{2}{10}$ | $\frac{3}{10}$ | | |
|---|---|---|---|---|---|

**2**

| $0$ | $\frac{2}{10}$ | $\frac{4}{10}$ | $\frac{6}{10}$ | | |
|---|---|---|---|---|---|

**3**

| $\frac{6}{10}$ | $\frac{8}{10}$ | $1$ | $1\frac{2}{10}$ | | |
|---|---|---|---|---|---|

**4**

| $0$ | $\frac{3}{10}$ | $\frac{6}{10}$ | $\frac{9}{10}$ | | |
|---|---|---|---|---|---|

**5**

| $1\frac{6}{10}$ | $1\frac{3}{10}$ | $1$ | $\frac{7}{10}$ | | |
|---|---|---|---|---|---|

**6**

| $1\frac{5}{10}$ | $1\frac{3}{10}$ | $1\frac{1}{10}$ | $\frac{9}{10}$ | | |
|---|---|---|---|---|---|

## SET 4 Extension

**1** $\frac{1}{4}$ of 56

**2** $8 \times 7 - 6$

**3** Share 98 stickers among 6.

**4** What numeral is in the thousands place in 7821?

**5** 3000 g = ☐ kg

**6** Write the numeral for seven thousand and ten.

**7** List the factors of 32.

**8** 5679 – 137

**9** How many edges has a cube?

**10** Is 36 a multiple of 9?

**11** How much is 7 kg of bacon at $3.50 per kilogram?

**12** Round 6328 to the nearest 100.

**13** Estimate an answer to $98 \times 5$.

**Mathematical Reasoning**

**14** Maria had $480 in the bank and Peter had half what Maria had plus $15. What was the total amount they had altogether?

# Measurement Displacement

How much water was displaced by each rock if there was 500 mL of water in each jug to begin?

**1**

☐ mL

**2**

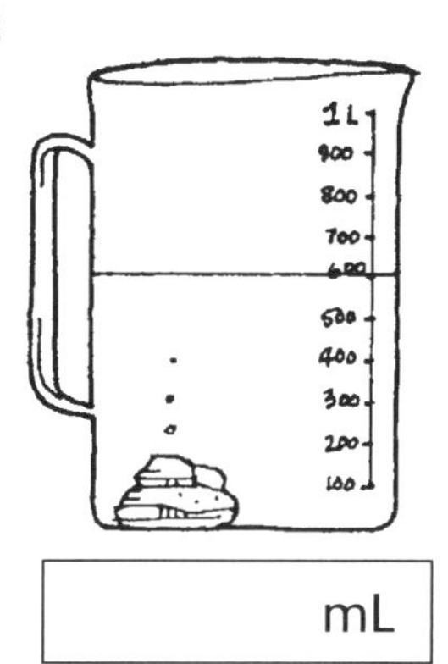

☐ mL

**3**

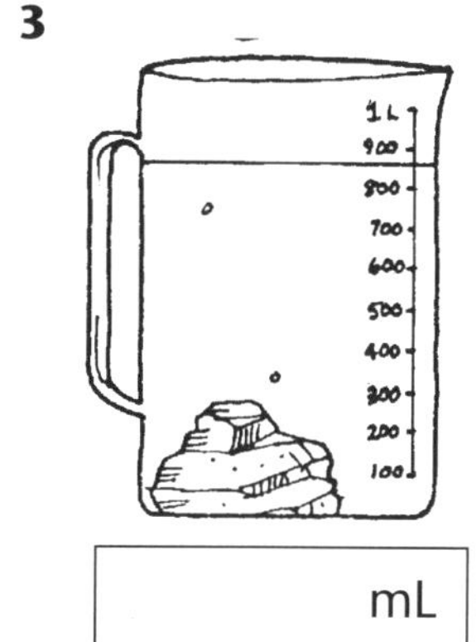

☐ mL

**4**

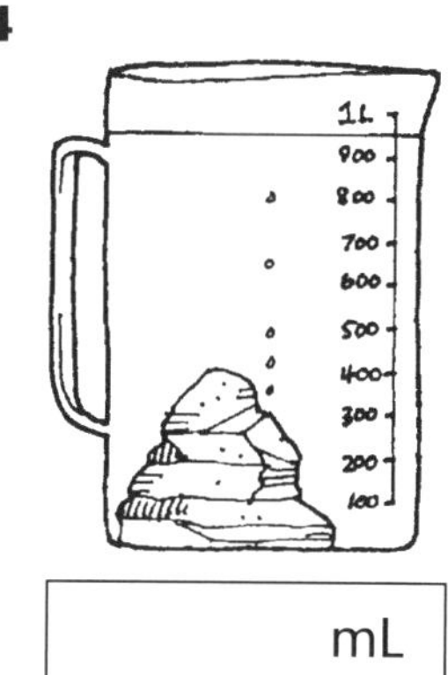

☐ mL

# Number and Algebra

## SET 1 Basic

1 $3 \times 5$

2 $48 - 5$

3 $17 + 8$

4 $32 \div 4$

5 $5 \times 8$

6 $20 \div 5$

7 What is the sum of 13 and 11?

8 What is the difference between 43 and 41?

9 Triple 8.

10 What is half of $42?

11 $16.03 = ☐ c

12 $6^2$

13

## SET 2 3-digit × 1-digit multiplication

Solve the multiplications.

1 $\begin{array}{r} 135 \\ \times \quad 3 \\ \hline \\ \hline \end{array}$

2 $\begin{array}{r} 146 \\ \times \quad 3 \\ \hline \\ \hline \end{array}$

3 $\begin{array}{r} 145 \\ \times \quad 4 \\ \hline \\ \hline \end{array}$

4 $\begin{array}{r} 167 \\ \times \quad 4 \\ \hline \\ \hline \end{array}$

5 $\begin{array}{r} 163 \\ \times \quad 5 \\ \hline \\ \hline \end{array}$

6 $\begin{array}{r} 149 \\ \times \quad 6 \\ \hline \\ \hline \end{array}$

7 6 paddocks had 107 cows in each. What was the total number of cows?

8 There were 5 boxes of books. If each box held 128 books, how many books were there altogether?

9 Helen had 8 boxes of mandarins. If there were 124 mandarins in each box, how many were there altogether?

# Space Combining shapes

Combine the squares and rectangles to form a pattern.

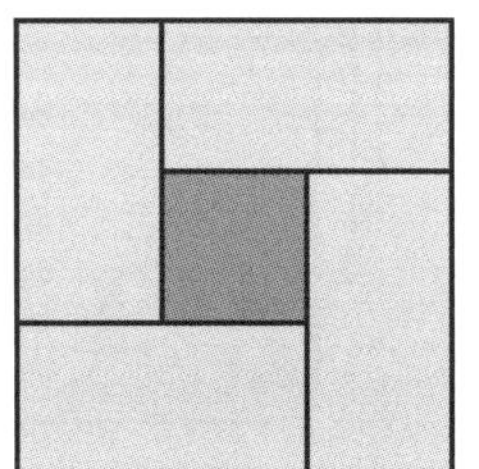

# Number and Algebra

## SET 3 Equivalence in equations

Use the = (equal to) or ≠ (not equal to) symbols to compare the groups.

| | | | |
|---|---|---|---|
| 1 | 17 + 16 + 13 | | 100 – 54 |
| 2 | 90 – 20 – 30 | | 23 + 19 + 17 |
| 3 | 8 × 5 | | 100 – 40 – 20 |
| 4 | 36 ÷ 4 | | 100 – 91 |
| 5 | 17 + 23 – 34 | | 40 ÷ 5 |
| 6 | 6 × 6 + 14 | | 5 × 8 + 30 |

Use the > (greater than), < (less than) or = (equal to) symbols to compare the number sentences.

| | | | |
|---|---|---|---|
| 7 | 80 + 30 – 7 | | 20 × 5 |
| 8 |  $\frac{1}{4}$ of 48 | | 36 ÷ 3 |
| 9 | 5 × 8 + 20 | | 18 + 12 + 47 |
| 10 | 3 × 9 + 13 | | 99 – 29 – 20 |

## SET 4 Extension

1 Write the number that is 5 before 303.

2 8 × 7 + 6

3 111, 131, 151, ☐

4 Subtract 70 from 3386.

5 How many corners has a triangular prism?

6 Write 8 tenths as a decimal.

7 How much are 9 apples at 45c each?

8 Write 4107 in words. ______________________

______________________

9 How many hours in 6 days?

10 Is 36 a multiple of 7?

11 How much is 4 $\frac{1}{2}$ kg of potatoes at 80c per kilogram?

12 Write the set of factors for 20.

13 Multiply $136 by 7.

14 4137, 4146, 4155, ☐

15 How many halves in 5 $\frac{1}{2}$?

16 The taxi made 9 trips. There were 45 passengers. What was the average number of passengers for each trip?

# Statistics and Probability Spreadsheets

| | A | B | C | D |
|---|---|---|---|---|
| 1 | Date | Item | Cost | Balance |
| 2 | 1 June | Opening | | $400 |
| 3 | 7 June | Food | $160 | (D2 – C3) $240 |
| 4 | 10 June | Petrol | $40 | (D3 – C4) $200 |
| 5 | 11 June | Bills | $120 | (D4 – C5) $80 |
| 6 | 20 June | Drinks | $30 | (D5 – C6) $50 |
| 7 | 29 June | Movies | $20 | (D6 – C7) $30 |

1 How much was spent on food?

2 How much was spent on bills?

3 How much was spent on movies?

4 What was the date when the balance equalled $200?

5 What was the date when the balance equalled $50?

6 Would there be enough money on 30 June to buy a radio costing $55?

UNIT
35

# Number and Algebra

## SET 1 Basic

1 $6 × 7

2 Share $32 among 8 people.

3 What is the difference between 20 and 9?

4 $16 × 2

5 What is the sum of 14, 18 and 20?

6 What is the difference between 90 and 39?

7 $6 + $7 + $7

8 $14.65 = ☐ c

9 What is the product of 7 and 8?

10 Triple 9.

11 13 × 3

12 15 kg – 9 kg

13 What is the difference between 48 and 30?

14 (2 + 1) × 2

15 90, 80, ☐, ☐, 50, 40, ☐

16 Jane had 40 marbles but lost 7 and gave away 13. How many did she have left? ☐

## SET 2 Decimal place value

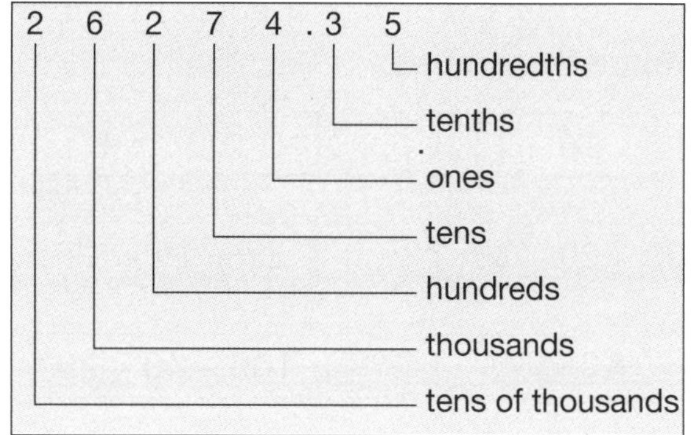


State the place value of each bold digit.

1 3**5**.21 ______

2 **2**16.57 ______

3 397.**6**8 ______

4 462.5**9** ______

5 3**5**74.62 ______

6 **6**987.43 ______

7 8549.6**4** ______

8 7412.**7**1 ______

9 2**6** 231.31 ______

10 **3**4 356.69 ______

11 5**6** 291.74 ______

# Statistics and Probability Posing questions

You have been given the task of improving the school playground.

You must survey some children to find out what improvements they would like.

Write 3 questions that you could ask in your survey.

1 ______

______

2 ______

______

3 ______

______

# Number and Algebra

## SET 3 Simple flow charts

Complete the flow charts.

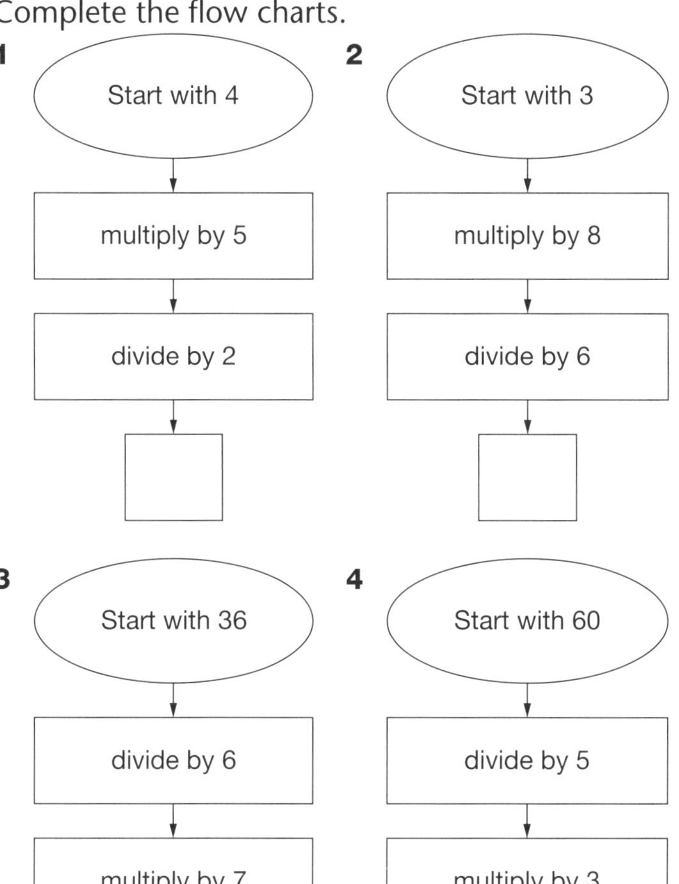


## SET 4 Extension

1 $4 \times 8 + 5$

2 How much are 10 pens at $4.50 each?

3 How many centimetres in $3\frac{3}{4}$ m?

4 Is 21 a multiple of 3?

5 How much would each person pay if 4 players hired a tennis court at $36 per court?

6 Multiply the sum of 5 and 4 by 7.

7 How much is $3\frac{1}{2}$ kg of meat at $7 per kilo?

8 Round off to the nearest 100 to estimate an answer to 698 + 717.

9 Lisa drank 5 mL of medicine at 8 am, then 5 mL every 4 hours. How much did she drink between 8 am and 9 pm? 

10 A rectangle has a perimeter of 34 m and an area of 72 $m^2$. What are the lengths of its sides?

Length ________ Width ________

Mathematical Reasoning

# Measurement Temperature

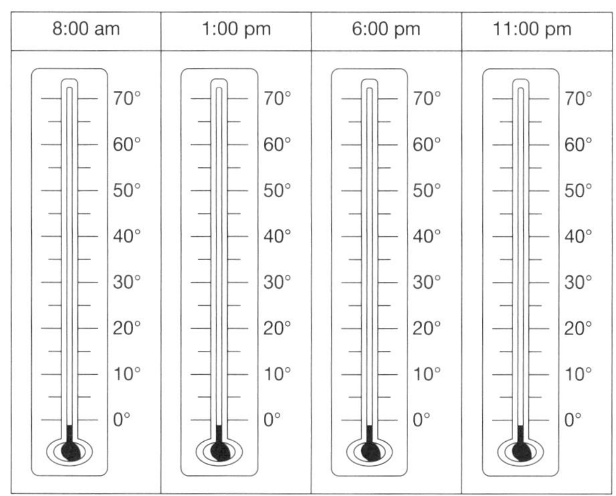


Record the temperatures on the thermometers.

1 At 8:00 am the temperature was 15°C.

2 By 1:00 pm it had risen 10°C.

3 By 6:00 pm it had fallen 5°C.

4 By 11:00 pm it had fallen another 10°C.

# Maths helpers

## Length

10 millimetres (mm) = 1 centimetre (cm)

100 centimetres (cm) = 1 metre (m)

1000 metres (m) = 1 kilometre (km)

## Mass

1000 grams (g) = 1 kilogram (kg)

1000 kg = 1 tonne (t)

## Capacity

1000 millilitres (mL) = 1 litre (L)

## Time

60 seconds = 1 minute

60 minutes = 1 hour

24 hours = 1 day

7 days = 1 week

14 days = 1 fortnight

12 months = 1 year

52 weeks = 1 year

365 days = 1 year

366 days = 1 leap year

10 years = 1 decade

100 years = 1 century

## Months of the year

Thirty days has September, April, June and November. All the rest have thirty-one, except February alone, which has twenty-eight days clear and twenty-nine days each leap year.

## Seasons

Summer: December, January, February

Autumn: March, April, May

Winter: June, July, August

Spring: September, October, November

## Roman numerals

1 = I

2 = II

3 = III

4 = IV

5 = V

6 = VI

7 = VII

8 = VIII

9 = IX

10 = X

20 = XX

30 = XXX

40 = XL

50 = L

60 = LX

70 = LXX

80 = LXXX

90 = XC

100 = C

500 = D

1000 = M

## Multiplication facts

| × | 0 | 1 | 2 | 3 | 4 | 5 | 6 | 7 | 8 | 9 | 10 |
|---|---|---|---|---|---|---|---|---|---|---|---|
| 0 | 0 | 0 | 0 | 0 | 0 | 0 | 0 | 0 | 0 | 0 | 0 |
| 1 | 0 | 1 | 2 | 3 | 4 | 5 | 6 | 7 | 8 | 9 | 10 |
| 2 | 0 | 2 | 4 | 6 | 8 | 10 | 12 | 14 | 16 | 18 | 20 |
| 3 | 0 | 3 | 6 | 9 | 12 | 15 | 18 | 21 | 24 | 27 | 30 |
| 4 | 0 | 4 | 8 | 12 | 16 | 20 | 24 | 28 | 32 | 36 | 40 |
| 5 | 0 | 5 | 10 | 15 | 20 | 25 | 30 | 35 | 40 | 45 | 50 |
| 6 | 0 | 6 | 12 | 18 | 24 | 30 | 36 | 42 | 48 | 54 | 60 |
| 7 | 0 | 7 | 14 | 21 | 28 | 35 | 42 | 49 | 56 | 63 | 70 |
| 8 | 0 | 8 | 16 | 24 | 32 | 40 | 48 | 56 | 64 | 72 | 80 |
| 9 | 0 | 9 | 18 | 27 | 36 | 45 | 54 | 63 | 72 | 81 | 90 |
| 10 | 0 | 10 | 20 | 30 | 40 | 50 | 60 | 70 | 80 | 90 | 100 |

## Addition facts

| + | 2 | 3 | 4 | 5 | 6 | 7 | 8 | 9 | 10 | 11 | 12 |
|---|---|---|---|---|---|---|---|---|---|---|---|
| 2 | 4 | 5 | 6 | 7 | 8 | 9 | 10 | 11 | 12 | 13 | 14 |
| 3 | 5 | 6 | 7 | 8 | 9 | 10 | 11 | 12 | 13 | 14 | 15 |
| 4 | 6 | 7 | 8 | 9 | 10 | 11 | 12 | 13 | 14 | 15 | 16 |
| 5 | 7 | 8 | 9 | 10 | 11 | 12 | 13 | 14 | 15 | 16 | 17 |
| 6 | 8 | 9 | 10 | 11 | 12 | 13 | 14 | 15 | 16 | 17 | 18 |
| 7 | 9 | 10 | 11 | 12 | 13 | 14 | 15 | 16 | 17 | 18 | 19 |
| 8 | 10 | 11 | 12 | 13 | 14 | 15 | 16 | 17 | 18 | 19 | 20 |
| 9 | 11 | 12 | 13 | 14 | 15 | 16 | 17 | 18 | 19 | 20 | 21 |
| 10 | 12 | 13 | 14 | 15 | 16 | 17 | 18 | 19 | 20 | 21 | 22 |
| 11 | 13 | 14 | 15 | 16 | 17 | 18 | 19 | 20 | 21 | 22 | 23 |
| 12 | 14 | 15 | 16 | 17 | 18 | 19 | 20 | 21 | 22 | 23 | 24 |

# Answers

## UNIT 1 Number and Algebra

### SET 1

1 12
2 17
3 4
4 15
5 30
6 28
7 20
8 March
9 14
10 3
11 6
12 7c
13 15
14 $7

### SET 2

1 71
2 55
3 77
4 99
5 102
6 108
7 161
8 201
9 True
10 False
11 True
12 True

### SET 3

1 6
2 9
3 3
4 15
5 30
6 12
7 18
8 24
9 27
10 21
11 3, 9, 15, 21, 27, 0, 18, 24
12 16 marbles

### SET 4

1 19
2 34
3 Sunday
4 50, 45, 40, 35, 30, 25
5 9
6 563c
7 7:45 pm
8 $1.15
9 $7.20
10 75c
11 9864
12 a $20, $20, $20, $10, $5
b $50, $10, $10, $5
c $50, $20, $5

### Space

My 2 bases are circles and my other face is curved. G — R PRISM
I have 6 faces which are all rectangles. R — B CONE
I have 1 base which is a circle. B — G CYLINDER
I have a square base and all my other faces are triangles. Y — Y PYRAMID

B G R Y

### Measurement

5 cm
7 cm
9 cm
13 cm
$14\frac{1}{2}$ cm
$15\frac{1}{2}$cm

## UNIT 2 Number and Algebra

### SET 1

1 9
2 10
3 3
4 2
5 15
6 24
7 20
8 30
9 January
10 9
11 5
12 7
13 60
14 46
15 5 shells

### SET 2

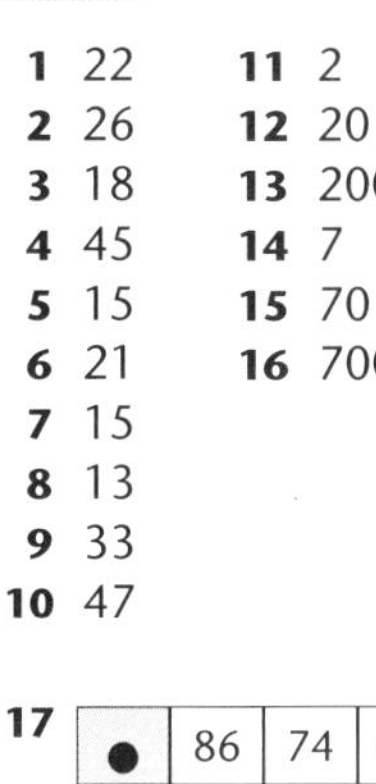

1 22
2 26
3 18
4 45
5 15
6 21
7 15
8 13
9 33
10 47
11 2
12 20
13 200
14 7
15 70
16 700

17

| | | | | | | |
|---|---|---|---|---|---|---|
| ● | 86 | 74 | 49 | 24 | 13 | 45 |
| ▲ | 77 | 65 | 40 | 15 | 4 | 36 |

18

| | | | | | | |
|---|---|---|---|---|---|---|
| ● | 35 | 55 | 85 | 95 | 125 | 135 |
| ▲ | 29 | 49 | 79 | 89 | 119 | 129 |

### SET 3

Hands on. Possible solutions

1 2 3

4 5 6

7 Shade 1 box
8 Shade 7 boxes
9 25 L, 20 L

### SET 4

1 C
2 10c each
3 24
4 16
5 $17.86
6 6 hundreds
7 726
8 22 days
9 $15
10 6
11 $10.50
12 987c
13 7 hrs
14 48
15 Seven thousand, two hundred and sixty
16 2

### Statistics and Probability

1 7
2 8
3 Toy cars
4 Books
5 28
6 Games and toy cars

### Measurement

| | Estimate | cm |
|---|---|---|
| 1 | Hands on | 10 cm |
| 2 | Hands on | 11 cm |
| 3 | Hands on | 12 cm |
| 4 | Hands on | 9 cm |
| 5 | Hands on | 7 cm |
| 6 | Hands on | 4 cm |

## UNIT 3 Number and Algebra

### SET 1

1 11
2 10
3 14
4 20
5 18
6 8
7 20
8 18
9 June
10 20
11 4
12 6
13 14
14 180
15 Five thousand, two hundred and seven

### SET 2

| | x | 3 | 2 | 5 | 1 | 0 | 7 | 9 |
|---|---|---|---|---|---|---|---|---|
| 1 | 3 | 9 | 6 | 15 | 3 | 0 | 21 | 27 |
| 2 | 0 | 0 | 0 | 0 | 0 | 0 | 0 | 0 |
| 3 | 4 | 12 | 8 | 20 | 4 | 0 | 28 | 36 |
| 4 | 10 | 30 | 20 | 50 | 10 | 0 | 70 | 90 |

5 $45
6 $18
7 $20
8 $30
9 $12
800 g
$18

### SET 3

1 10, 10
2 17, 17
3 90, 90
4 150, 150
5 132, 132
6 276, 276
7 No
8 30, 30
9 28, 28
10 24, 24
11 48, 48
12 35, 35
13 32, 32
14 No

### SET 4

1 E
2 5
3 20c each
4 67c
5 61
6 6 thousands
7 1500
8 $4.50
9 4000
10 No
11 20
12 $170
13 180 min
14 35
15 Three thousand and sixty-two
16 45

### Space

Hands on. Some examples below:

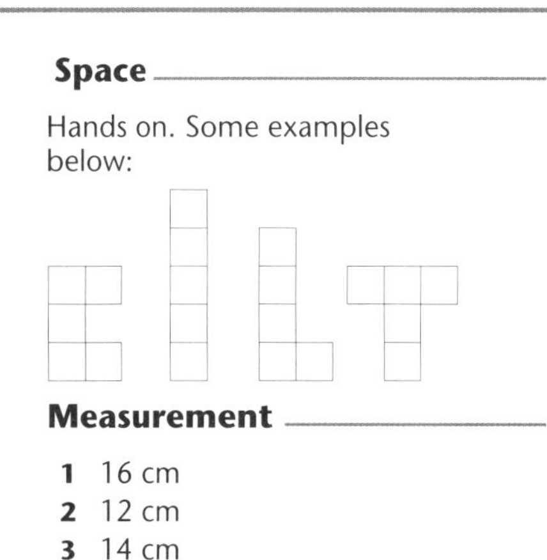

### Measurement

1 16 cm
2 12 cm
3 14 cm
4 18 cm

# Answers

## UNIT 4 Number and Algebra

SET 1

1 14
2 16
3 13
4 11
5 40
6 45
7 21
8 12
9 33
10 37
11 18
12 April
13 30 min
14 27
15 $3237

SET 2

1 seven hundreds
2 four tens
3 three thousands
4 six ones
5 five hundreds
6 856
7 2457
8 4576
9 2775
10 3474
11 397
12 1397
13 3540
14 3464
15 1296
16 753
17 357

SET 3

1 $\frac{2}{5}$
2 $\frac{7}{10}$
3 $\frac{3}{10}$
4 $\frac{4}{5}$
5 $\frac{4}{10}$
6 $\frac{4}{5}$
7 $\frac{1}{10}, \frac{2}{10}, \frac{5}{10}, \frac{9}{10}$
8 5
9 Yes
10 Yes
11 Hands on

SET 4

1 H
2 500 g
3 $8 each
4 $9
5 $4.50
6 8 tens
7 70
8 280
9 No (5)
10 $21
11 Possible answers include:
50 + 30 + 20 = 100
50 + 10 + 40 = 100
40 + 40 + 20 = 100

### Space

1 Yes
2 No
3 Yes
4 Yes
5 No
6 Yes

### Statistics and Probability

1 4
2 8
3 11
4 6
5 6

## UNIT 5 Number and Algebra

SET 1

1 7
2 3
3 8
4 4
5 20
6 6
7 24
8 4
9 February
10 18
11 22
12 22
13 120 min
14 4 thousands, 3 hundreds, 7 tens, 2 ones
15 11 km

SET 2

1 642
2 534
3 934
4 870
5 150
6 130
7 230
8 340
9 450
10 435
11 550

SET 3

1 12
2 120
3 16
4 160
5 18
6 180
7 21
8 210
9 16
10 160
11 28
12 280
13 25
14 250
15 36
16 360
17 30 mL
18 60 mL

SET 4

1 95
2 85
3 100
4 20
5 9
6 30c
7 950
8 22
9 12
10 5
11 4
12 2640
13 48
14 4085
15 1324
16



### Space

### Measurement

1 10 squares
2 14 squares
3 18 squares

## UNIT 6 Number and Algebra

SET 1

1 29
2 80
3 15
4 21
5 24
6 12
7 15
8 16
9 15
10 18
11 13
12 9
13 24
14 8 hrs
15 7637
16 $18

SET 2

1 8, 16, 20, 24, 28, 32
2 10, 15, 20, 30, 40, 45
3 12, 24, 30, 36, 54, 48
4 $24
5 $21
6 $48
7 $40
8 14
9 30
10 18
11 180
12 12
13 120
14 35
15 350

SET 3

1 64
2 99
3 188
4 198
5 198
6 278
7 379
8 491
9 582
10 702
11 395
12 494
13 $587

SET 4

1 35c
2 40
3 $2.20
4 9 hundreds
5 About 1800
6 1590
7 1600
8 3000
9 12 hrs
10 29
11 9, 18, 27, 36, 45, 54, 63
12 $1.35
13 6699
14 $240

### Space

| yellow | blue | red | blue | yellow |
|---|---|---|---|---|
| blue | yellow | blue | yellow | red |

### Measurement

1 A bucket full of water
2 A TV set
3 1 kg mass
4 2 kg of potatoes
5 1 kg mass
6 1 kg bag of feathers
7 1 kg

## UNIT 7 Number and Algebra

**SET 1**

1 19
2 2
3 8
4 10
5 14
6 24
7 7
8 18
9 5
10 63
11 9
12 15
13 35
14 18
15 $13

**SET 2**

1 21, 21
2 170, 170
3 220, 220
4 30, 30
5 40, 40
6 30, 30
7 169
8 129
9 60
10 200

**SET 3**

Hands on. Some examples

1
2
3
4
5 False
6 True
7 False
8 True

**SET 4**

1 120
2 331
3 5 hundreds
4 2800
5 30
6 164
7 $7 each
8 14
9 315c
10 16
11 40
12 53
13 3 at $1.10 each and $1 + $1.20 + $1.10

### Statistics and Probability

1 Walk
2 Bike
3 50
4 Bike and train

### Measurement

Hands on

## UNIT 8 Number and Algebra

**SET 1**

1 13
2 10
3 16
4 16 km
5 18
6 28
7 20
8 0
9 May
10 19
11 $2 each
12 24
13 10c
14 4:18
15 643
16 895c

**SET 2**

1 426
2 489
3 371
4 354
5 300
6 1100
7 420
8 827
9 921
10 7046
11 $441

**SET 3**

Missing Numbers:
1 20, 25, 30, 35, 40
2 9, 12, 15, 18, 21
3 50, 60, 70, 80, 90
4 32, 35, 38, 41, 44
5 222, 224, 226, 228, 230
6 310, 305, 300, 295, 290
7 350, 300, 250, 200, 150
8 700, 750, 800, 850, 900
9 845
10 4231
11 3707

**SET 4**

1 127 cm
2 F
3 6
4 9 each
5 2000 g
6 2 hundreds
7 62
8 3206
9 2497
10 1300
11 1378
12 $12
13 12 hrs
14 28
15 One thousand, four hundred
16 $2
17 About 250

### Number and Algebra

1 20
2 30
3 30
4 50
5 70
6 90
7 20 + 20 = 40
8 40 + 30 = 70
9 50 + 50 = 100
10 90 + 60 = 150
11 100 + 80 = 180

### Measurement

1 milk carton
2 cordial
3 4 L juice
4 bucket
5 drum

## UNIT 9 Number and Algebra

**SET 1**

1 21
2 20
3 2
4 18
5 20
6 33
7 32
8 20
9 July
10 48
11 4 each
12 25
13 3
14 1:25 pm
15 $8

**SET 2**

1 6, 12, 15, 18, 21, 24
2 14, 21, 28, 42, 56, 63
3 10, 20, 25, 30, 45, 40
4 $42
5 $21
6 $48
7 $56
8 41
9 85
10 70
11 42
12 45
13 70
14 21
15 210

**SET 3**

1 [4000] + [700] + [30] + [6]
2 [5000] + [600] + [70] + [4]
3 1233, 1235
4 5677, 5679
5 2852, 2854
6 1299, 44 974, 73 567
7 13 679, 24 674, 35 675
8 44 376, 57 643, 66 743
9 8060
10 3588
11 2401
12 6070
13 1556

**SET 4**

1 $9.90
2 48
3 270
4 10
5 46
6 $22.95
7 80
8 148
9 24
10 1390
11 16
12 50c each
13 $4
14 72
15 One thousand, nine hundred and sixty-four
16 32

### Space

### Statistics and Probability

Hands on

# Answers

## UNIT 10 Number and Algebra

**SET 1**

1 13 cm
2 1
3 32
4 20c
5 30
6 12
7 14
8 32
9 October
10 22
11 28
12 18
13 32
14 9 hrs
15 7137
16 9

**SET 2**

1 28
2 36
3 44
4 84
5 160
6 120
7 70
8 110
9 130
10 90
11 230
12 120
13 72
14 55
15 192
16 96

**SET 3**

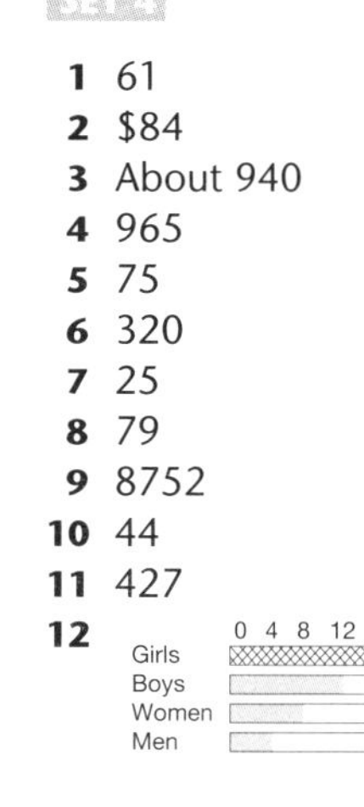


5 True
6 False
7 False
8 True
9 True
10 False

**SET 4**

1 61
2 $84
3 About 940
4 965
5 75
6 320
7 25
8 79
9 8752
10 44
11 427
12



### Space

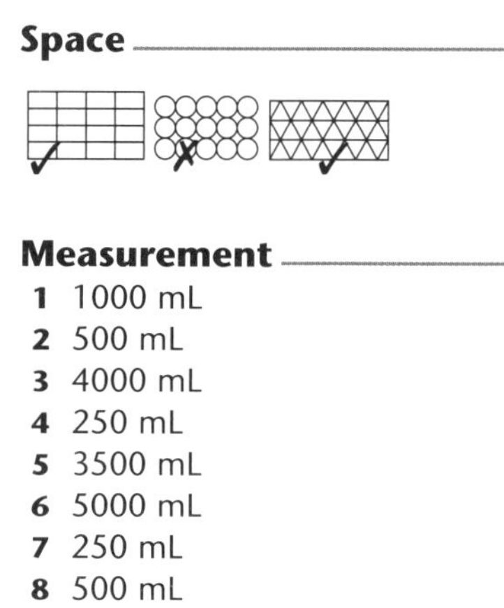

### Measurement

1 1000 mL
2 500 mL
3 4000 mL
4 250 mL
5 3500 mL
6 5000 mL
7 250 mL
8 500 mL

## UNIT 11 Number and Algebra

**SET 1**

1 16
2 58
3 8
4 11
5 12
6 0
7 40
8 42
9 6
10 45
11 14
12 4
13 15
14 4 o'clock
15 963

**SET 2**

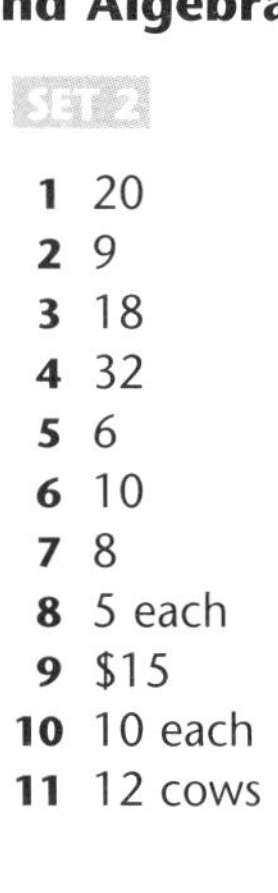

1 20
2 9
3 18
4 32
5 6
6 10
7 8
8 5 each
9 $15
10 10 each
11 12 cows

**SET 3**

1 24
2 40
3 56
4 72
5 80
6 48
7 32
8 88
9 16
10 96
11
12
13
14

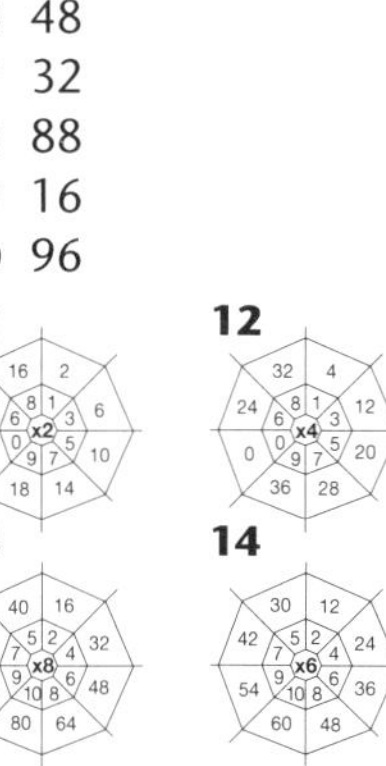

**SET 4**

1 368
2 956
3 7631
4 $25 each
5 28
6 7000
7 3000
8 7026
9 243
10 2
11 270 min
12 $56
13 $36
14 $20

### Statistics and Probability

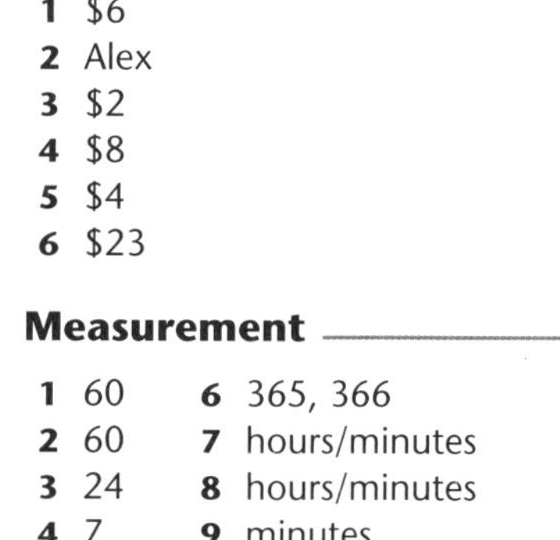

1 $6
2 Alex
3 $2
4 $8
5 $4
6 $23

### Measurement

1 60
2 60
3 24
4 7
5 14
6 365, 366
7 hours/minutes
8 hours/minutes
9 minutes
10 seconds

## UNIT 12 Number and Algebra

**SET 1**

1 15
2 24
3 14
4 29
5 21
6 12
7 70
8 36c
9 19 kg
10 22
11 15
12 30
13 19
14 4
15 639
16 $8

**SET 2**

1 215
2 108
3 228
4 125
5 124
6 112
7 180
8 168
9 222
10 126
11 120
12 200
13 490

**SET 3**

Possible solutions:

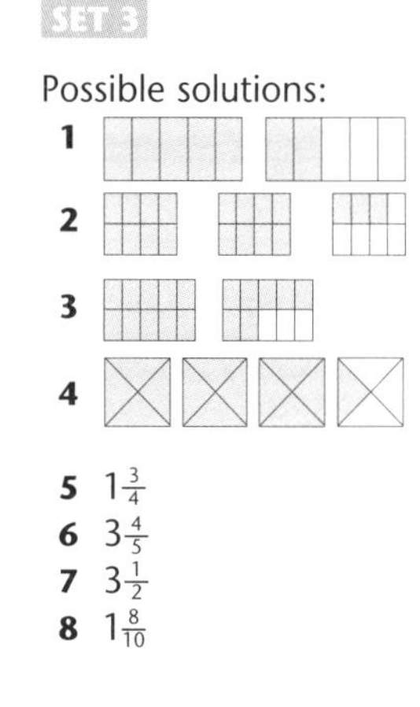

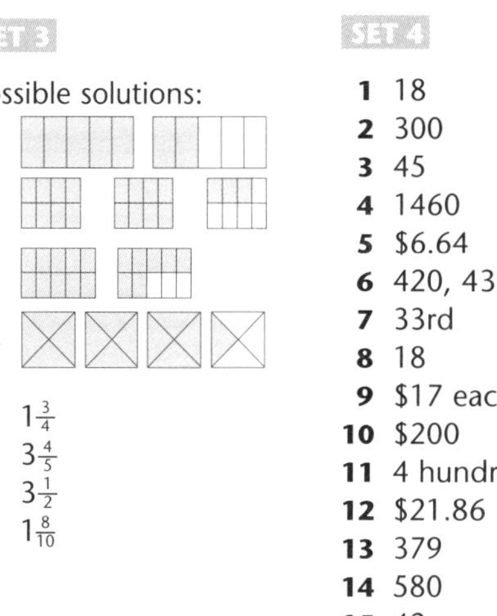

1
2
3
4

5 $1\frac{3}{4}$
6 $3\frac{4}{5}$
7 $3\frac{1}{2}$
8 $1\frac{8}{10}$

**SET 4**

1 18
2 300
3 45
4 1460
5 $6.64
6 420, 435
7 33rd
8 18
9 $17 each
10 $200
11 4 hundreds
12 $21.86
13 379
14 580
15 42
16 93
17 12

### Space

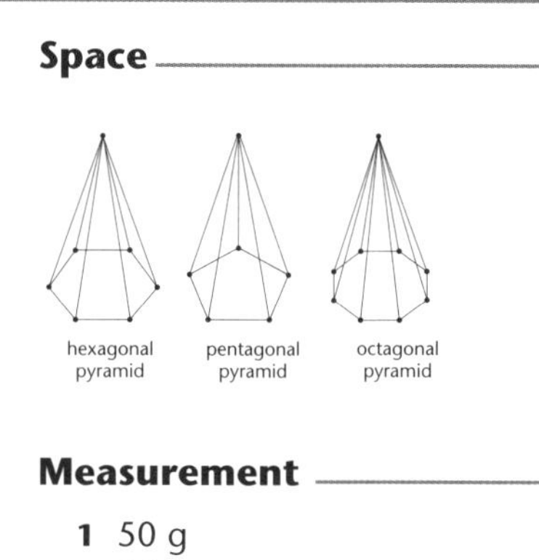


### Measurement

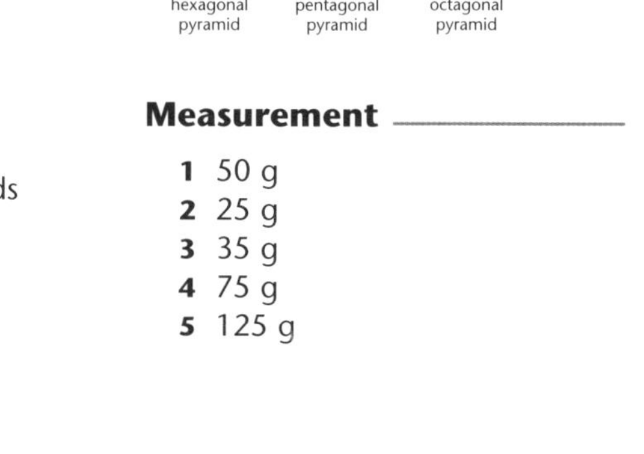

1 50 g
2 25 g
3 35 g
4 75 g
5 125 g

## UNIT 13 Number and Algebra

### SET 1

1 9
2 12
3 28
4 30
5 32
6 48
7 24
8 21c
9 31
10 386c
11 7
12 5
13 25
14 1 thousand, 3 hundreds, 7 tens, 6 ones
15 926

### SET 2

1 281
2 375
3 468
4 585
5 430
6 680
7 782
8 12 / 18 6 24 / 30

### SET 3

| | Number | Nearest 100 |
|---|---|---|
| 1 | 976 | 1000 |
| 2 | 823 | 800 |
| 3 | 1213 | 1200 |
| 4 | 1787 | 1800 |
| 5 | 2356 | 2400 |
| 6 | 7599 | 7600 |
| 7 | 4207 | 4200 |

8 1000
9 1000
10 1300
11 1900
12 1700
13 2000

### SET 4

1 32
2 5 each, remainder 1
3 4825
4 $1.80
5 6 hundreds
6 17
7 $6.10
8 $135
9 6
10 12
11 $28
12 400
13 $7 each
14 402, 399, 396, 393, 390, 387
15 5

### Measurement

1 m
2 cm
3 m
4 cm
5 m
6 200 cm
7 500 cm
8 900 cm
9 150 cm
10 125 cm

### Statistics and Probability

1 A
2 C
3 C
4 B
5 A

## UNIT 14 Number and Algebra

### SET 1

1 22
2 20
3 12
4 32cm
5 21
6 15
7 21
8 50c
9 25 kg
10 15
11 21
12 24
13 11 hours
14 28
15 $11.00

### SET 2

1 5874
2 8704
3 8622
4 6932
5 4429
6 9545
7 $\begin{array}{r} 345\mathbf{3} \\ +\,2\mathbf{8}36 \\ \hline 6289 \end{array}$
8 $\begin{array}{r} 2\mathbf{7}46 \\ +\,35\mathbf{72} \\ \hline 6318 \end{array}$

### SET 3

1 6
2 4
3 3
4 4
5 2
6 4
7 10 each
8 9 each
9 6 each
10 3
11 9
12 4
13 3
14 7
15 9

### SET 4

1 9620
2 25
3 210 min
4 $7 each
5 358, 367
6 $40
7 $5.75
8 436th
9 16
10 1, 2, 3, 4, 6, 8, 12, 24
11 $48
12 52
13 330
14 $8 \times 3 \times 2 = 48$ or $8 \times 2 \times 3 = 48$

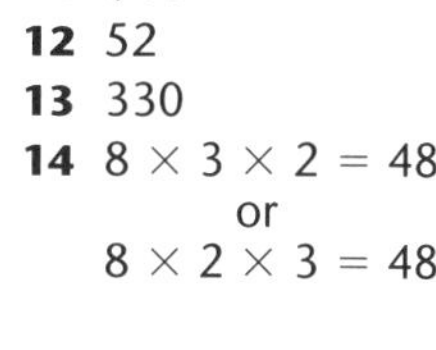

### Number and Algebra

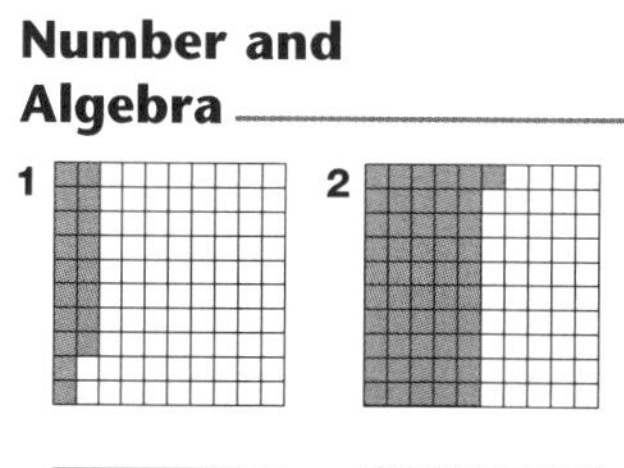

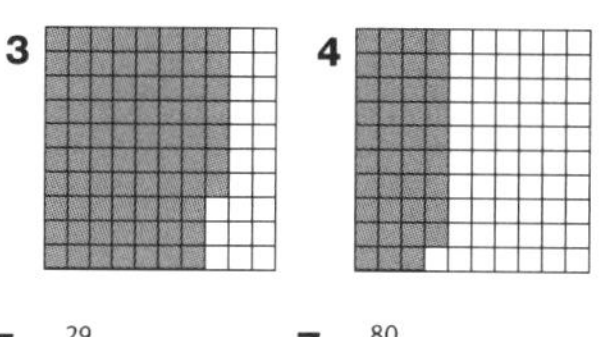

5 $\frac{29}{100}$
6 $\frac{45}{100}$
7 $\frac{80}{100}$
8 $\frac{92}{100}$

### Measurement

1 14 $cm^2$
2 Hands on

## UNIT 15 Number and Algebra

### SET 1

1 22
2 42
3 14
4 $24
5 9
6 16
7 40c
8 88
9 55 kg
10 44
11 18
12 10
13 15
14 3:55 pm
15 6 hundreds, 9 tens, 5 ones
16 15c

### SET 2

1 18
2 27
3 9
4 45
5 90
6 36
7 54
8 72
9 81
10 63
11 7, 21, 35, 49 63, 0, 42, 56
12 5, 15, 25, 35, 45, 0, 30, 40

### SET 3

1 $1\frac{1}{4}, 1\frac{2}{4}$
2 $3, 3\frac{1}{2}$
3 $1\frac{1}{5}, 1\frac{2}{5}$
4 $1\frac{1}{8}, 1\frac{2}{8}$
5 $\frac{1}{10}, \frac{3}{10}, \frac{7}{10}$
6 $\frac{1}{4}, \frac{2}{4}, \frac{3}{4}$
7 $\frac{3}{8}, \frac{5}{8}, \frac{7}{8}$
8 $1\frac{1}{4}, 1\frac{2}{4}, 1\frac{3}{4}$,
9 $2\frac{1}{5}, 2\frac{2}{5}, 2\frac{3}{5}$
10 True
11 False
12 True
13 False

### SET 4

1 $23.80
2 $4.60
3 $9.75
4 16th
5 42
6 54, 72, 90, 108
7 $1.32
8 $1.55
9 1, 2, 3, 4, 6, 9, 12, 18, 36
10 350 cm
11 Green Point, 26
12 580
13 9

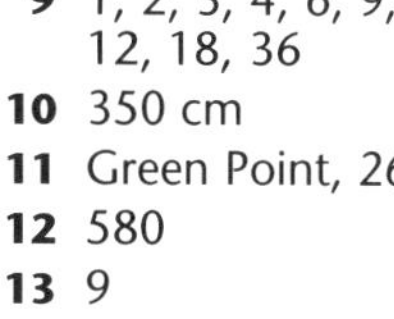

### Space

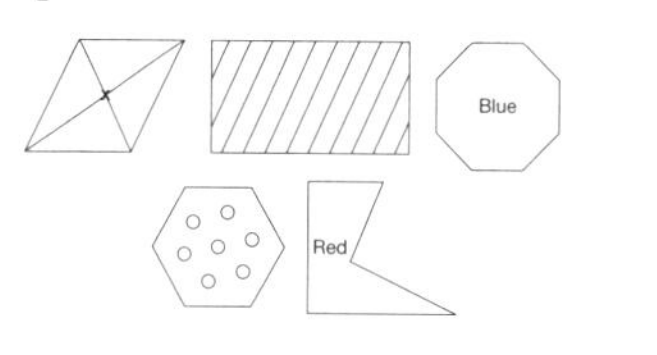


### Measurement

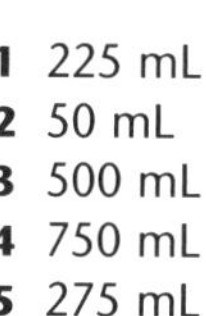

1 225 mL
2 50 mL
3 500 mL
4 750 mL
5 275 mL
6 700 mL

# Answers

## UNIT 16 Number and Algebra

### SET 1

1 42
2 37
3 33
4 43
5 28
6 18
7 35
8 48
9 22
10 10
11 14
12 4
13 63
14 18
15 $2.60

### SET 2

1 5312
2 2424
3 3479
4 4907
5 1215 m
6 884 m

### SET 3

1 No
2 No
3 Yes
4 No
5 No
6 No
7 Yes

### SET 4

1 317 cm
2 265
3 450 cm
4 159
5 6 hundreds
6 $18
7 208
8 359
9 190
10 24
11 $54
12 41
13 37
14 8
15 Four thousand, two hundred and seven
16 1, 2, 3, 4, 6, 8, 12, 16, 24, 48
17 0.3
18 3900

### Statistics and Probability

1 unlikely
2 unlikely
3 likely
4 impossible

### Measurement

1 1 kg 600 g
2 2 kg 260 g
3 4 kg 950 g
4 5 kg 20 g
5 3 kg 35 g

## UNIT 17 Number and Algebra

### SET 1

1 47
2 89
3 22
4 41c
5 42
6 21
7 28
8 35
9 19
10 12
11 20
12 893c
13 3
14 64
15 $39

### SET 2

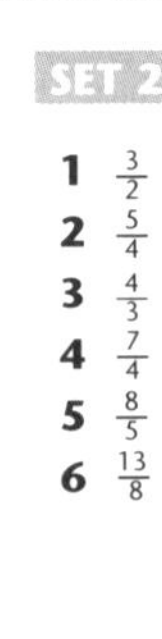

1 $\frac{3}{2}$
2 $\frac{5}{4}$
3 $\frac{4}{3}$
4 $\frac{7}{4}$
5 $\frac{8}{5}$
6 $\frac{13}{8}$

### SET 3

1

| Hours | 1 | 2 | 3 | 4 | 5 | 6 |
|---|---|---|---|---|---|---|
| Pay | $8 | $16 | $24 | $32 | $40 | $48 |

2

| Hours | 1 | 2 | 3 | 4 | 5 | 6 |
|---|---|---|---|---|---|---|
| Kilometres | 4 | 8 | 12 | 16 | 20 | 24 |

3

| Hours | 1 | 2 | 3 | 4 | 5 | 6 |
|---|---|---|---|---|---|---|
| Pay | $7 | $14 | $21 | $28 | $35 | $42 |

4

| Hours | 1 | 2 | 3 | 4 | 5 | 6 |
|---|---|---|---|---|---|---|
| Kilograms | 9 | 18 | 27 | 36 | 45 | 54 |

5

| Kilograms | 1 | 2 | 3 | 4 | 5 | 6 |
|---|---|---|---|---|---|---|
| Tomatoes | 6 | 12 | 18 | 24 | 30 | 36 |

### SET 4

1 36
2 257
3 About 550
4 10
5 $14.95
6 $3.10
7 1824
8 $33.50
9 350
10 $4.30
11 1327, 1427
12 $14
13 44
14 $129
15 Two thousand, four hundred and ninety-five
16 9750
17 No

### Space

### Measurement

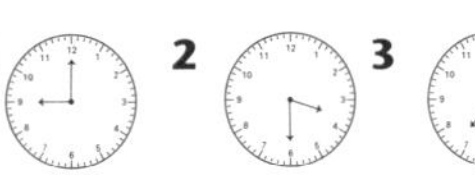

1 2 3

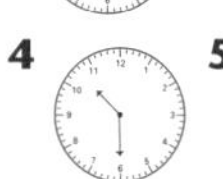

4 5 6

## UNIT 18 Number and Algebra

### SET 1

1 14
2 35
3 19
4 $14
5 16
6 14
7 70c
8 89
9 63 kg
10 72
11 29
12 20
13 14
14 9 hundreds, 6 tens, 2 ones
15 3:22

### SET 2

1 9841
2 9732
3 5169
4 3610
5 $2178
6 $4930
7 $3639

### SET 3

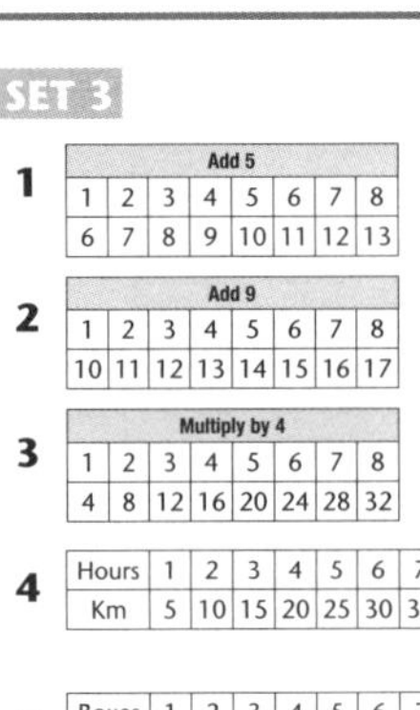

1

| Add 5 | | | | | | | |
|---|---|---|---|---|---|---|---|
| 1 | 2 | 3 | 4 | 5 | 6 | 7 | 8 |
| 6 | 7 | 8 | 9 | 10 | 11 | 12 | 13 |

2

| Add 9 | | | | | | | |
|---|---|---|---|---|---|---|---|
| 1 | 2 | 3 | 4 | 5 | 6 | 7 | 8 |
| 10 | 11 | 12 | 13 | 14 | 15 | 16 | 17 |

3

| Multiply by 4 | | | | | | | |
|---|---|---|---|---|---|---|---|
| 1 | 2 | 3 | 4 | 5 | 6 | 7 | 8 |
| 4 | 8 | 12 | 16 | 20 | 24 | 28 | 32 |

4

| Hours | 1 | 2 | 3 | 4 | 5 | 6 | 7 |
|---|---|---|---|---|---|---|---|
| Km | 5 | 10 | 15 | 20 | 25 | 30 | 35 |

5

| Boxes | 1 | 2 | 3 | 4 | 5 | 6 | 7 |
|---|---|---|---|---|---|---|---|
| Cans | 8 | 16 | 24 | 32 | 40 | 48 | 56 |

### SET 4

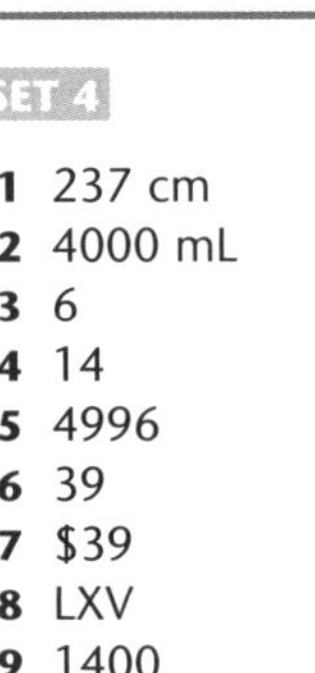

1 237 cm
2 4000 mL
3 6
4 14
5 4996
6 39
7 $39
8 LXV
9 1400
10 5
11 42
12 4

### Measurement

1 C
2 B
3 D

### Statistics and Probability

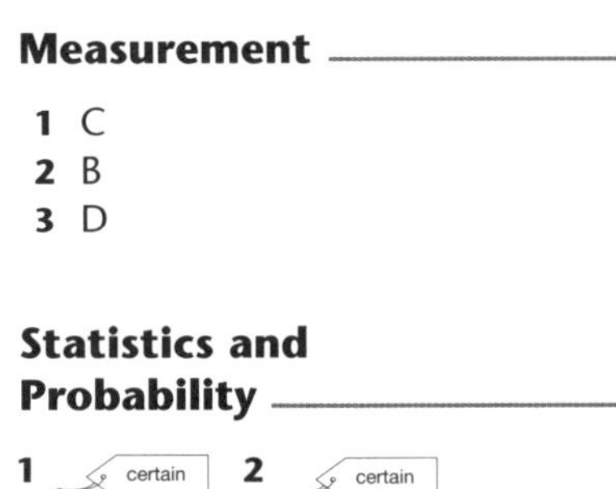

## UNIT 19 Number and Algebra

**SET 1**

1 16
2 13
3 42
4 28
5 56
6 49
7 62
8 46
9 20
10 18
11 586c
12 13
13 56
14 7
15 50c

**SET 2**

1 25 – 6 = 19 or
25 – 19 = 6
2 13 + 12 = 25
3 42 ÷ 6 = 7 or
42 ÷ 7 = 6
4 9 × 5 = 45 or
5 × 9 = 45
5 6 × 6 = 36
6 and 7

6

| | THOU | HUND | TENS | ONES |
|---|---|---|---|---|
| | 9 | 1 | 9 | 3 |
| – | | 4 | 2 | 5 |
| | 8 | 7 | 6 | 8 |

7

| | THOU | HUND | TENS | ONES |
|---|---|---|---|---|
| | 8 | 0 | 3 | 3 |
| – | 2 | 3 | 4 | 5 |
| | 5 | 6 | 8 | 8 |

**SET 3**

1 Shade any 35 squares.
2 Shade any 47 squares.
3 Shade any 96 squares.
4 Shade any 83 squares.
5 0.27
6 0.54
7 0.09, 0.29, 0.35
8 0.29, 0.33, 0.39
9 0.87, 1.07, 7.01
10 8.59, 9.58, 9.85

**SET 4**

1 483
2 4 m 37 cm
3 18
4 3500 kg
5 2 hrs
6 7
7 $6 each
8 8546
9 42
10 54
11 About 390
12 3999
13 $5
14 Length = 40 cm
Width = 20 cm

### Measurement

1 9°C
2 5°C
3 4°C
4 14°C
5 1°C
6 22°C

### Measurement

1

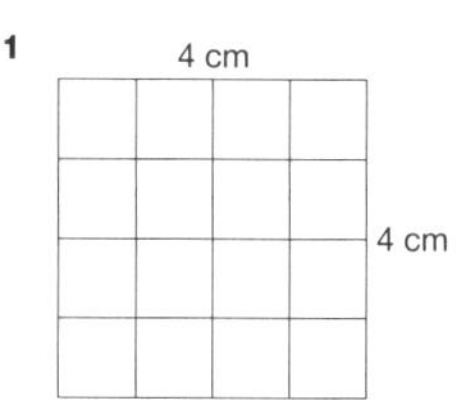


2

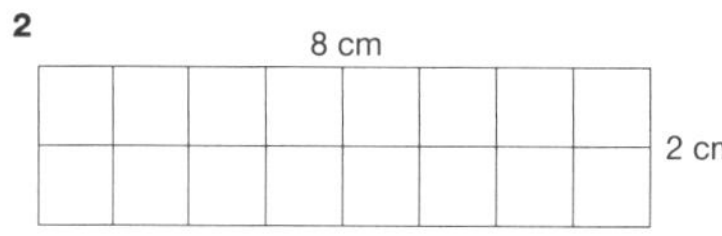


## UNIT 20 Number and Algebra

**SET 1**

1 16
2 18
3 10
4 29
5 21c
6 21
7 927c
8 3
9 29
10 420
11 $7 each
12 $88
13 56 kg
14 Two thousand and sixty

**SET 2**

1 6
2 4
3 4
4 5
5 5
6 7
7 8
8 8
9 9
10 12
11 Hands on
12 Hands on

**SET 3**

1 0.1
2 0.7
3 0.9
4 0.3
5 0.5
6 0.8
7 0.4
8 0.6
9-12 Hands on. Some examples below:

9 10

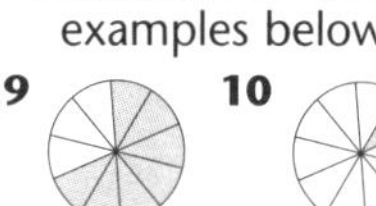


11

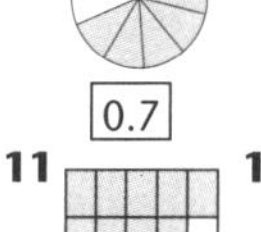


12

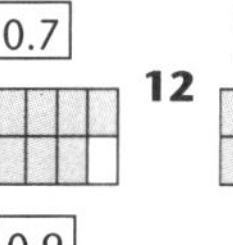


13 Complete the pattern

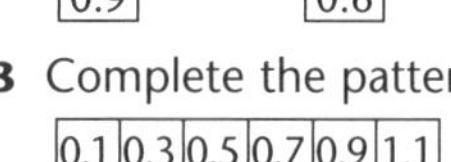


**SET 4**

1 32
2 1258
3 8
4 1706c
5 $24.20
6 $3.05
7 50
8 $10.95
9 140
10 $1.17
11 $7.50
12 $77
13 9950
14 7 m
15 28 cm
16 7

### Statistics and Probability

1

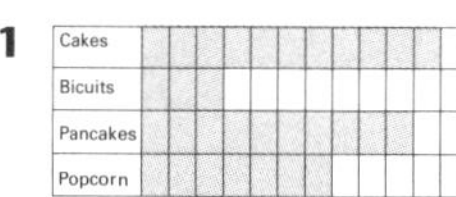


2 cakes
3 31

### Measurement

1 18 cm
2 14 cm
3 28.6 m

## UNIT 21 Number and Algebra

**SET 1**

1 10
2 8
3 18
4 36
5 14
6 48
7 36
8 21
9 21
10 17
11 $3.70
12 6
13 25
14 5
15 4 thousands, 0 hundreds, 5 tens, 7 ones

**SET 2**

1 5 × 5 = 25
2 8 × 4 = 32 or
4 × 8 = 32
3 4 × 10 = 40 or
10 × 4 = 40
4 4 × 7 = 28 or
7 × 4 = 28
5 9 × 9 = 81
6 6 × 7 = 42 or
7 × 6 = 42
7 6 × 6 = 36
8 9 × 6 = 54 or
6 × 9 = 54
9 7 r 1
10 3 r 2

**SET 3**

| | Tenths | Hundredths | Decimal |
|---|---|---|---|
| 1 | $\frac{1}{10}$ | $\frac{10}{100}$ | 0.10 |
| 2 | $\frac{3}{10}$ | $\frac{30}{100}$ | 0.30 |
| 3 | $\frac{9}{10}$ | $\frac{90}{100}$ | 0.90 |
| 4 | $\frac{5}{10}$ | $\frac{50}{100}$ | 0.50 |
| 5 | $\frac{7}{10}$ | $\frac{70}{100}$ | .70 |
| 6 | $\frac{2}{10}$ | $\frac{20}{100}$ | 0.20 |
| 7 | $\frac{6}{10}$ | $\frac{60}{100}$ | 0.60 |

8 True
9 True
10 True
11 False
12 True
13 True
14 False
15 False
16 True

**SET 4**

1 3000 g
2 337
3 405 cm
4 36
5 1 hundred
6 12
7 128
8 1839
9 $16
10 92
11 $5.50
12 $3.25
13 7 r 2
14 100 cm and 80 cm

### Space

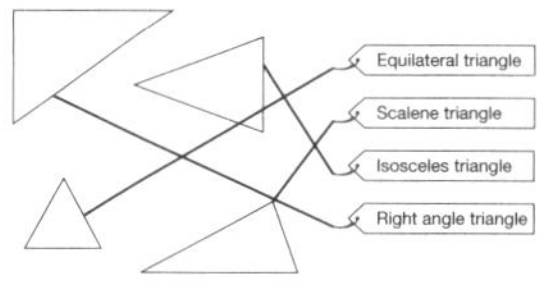


### Measurement

1 5:00 am
2 10:15 pm
3 12:25 pm
4 11:40 pm
5 5:45 am
6 1:30 pm

# Answers

## UNIT 22 Number and Algebra

### SET 1

1 35
2 15
3 3
4 35
5 8
6 10
7 30
8 24
9 3
10 105
11 3
12 407c
13 14
14 3 thousands, 2 hundreds, 0 tens, 4 ones
15 Two thousand and five

### SET 2

1 tenths
2 hundredths
3 tens
4 ones
5 hundredths
6 tenths
7 hundreds
8 tenths
9 0.53, 0.67, 0.76
10 0.93, 1.07, 1.86
11 1.24, 1.42, 12.03

### SET 3

1 18
2 17
3 26
4 16
5 18
6 15
7 13
8 17
9 13
10 14
11 12
12 18
13 6
14 15
15 16

### SET 4

1 10
2 19
3 $33
4 $2471
5 8 each
6 64
7 4
8 4837
9 5 tens
10 432nd
11 $8.35
12 143
13 15
14 48
15 Four thousand, three hundred and nine
16 $3.60

### Space

1 70 m
2 40 m
3 70 m
4 20 m

### Measurement

1 6 cubes
2 12 cubes
3 12 cubes
4 8 cubes

## UNIT 23 Number and Algebra

### SET 1

1 48
2 9
3 25
4 10
5 40
6 19
7 34
8 25
9 80
10 9, 15, 18
11 5
12 40
13 V
14 202c
15 8 thousands, 6 hundreds, 9 tens, 4 ones
16 $35.05

### SET 2

1 $180
2 500
3 $354
4 $102
5 $68
6 $277

### SET 3

1 0.5 | 0.7 | 0.2
2 0.6 | 0.9 | 0.3
3 0.34 | 0.58 | 0.41
4 2.30 | 2.75 | 2.12
5 3.12 m
6 4.1 m

### SET 4

1 3187
2 4000 g
3 2 m 47 cm
4 4
5 35 cm
6 6
7 $6
8 $4 each
9 14
10 $3.15
11 $1.80
12 5:30
13 Yes
14 35
15 Four hundred and ninety-seven
16 25 min
17 3700

### Number and Algebra

| | Date | purchase | Debit | Balance |
|---|---|---|---|---|
| 1 | 1/10 | Opening | - | $198.00 |
| 2 | 2/10 | Cheergirl | $17.00 | $181.00 |
| 3 | 3/10 | Bella's | $11.00 | $170.00 |
| 4 | 4/10 | Style Mart | $20.00 | $150.00 |
| 5 | 5/10 | Walkaway | $35.00 | $115.00 |
| 6 | 6/10 | Musica | $25.00 | $90.00 |
| 7 | 7/10 | Best Bags | $45.00 | $45.00 |

### Measurement

1 1 kg 400 g
2 2 kg 800 g
3 1 kg 800 g
4 2 kg 300 g

## UNIT 24 Number and Algebra

### SET 1

1 24
2 8
3 38
4 5
5 15
6 32
7 62
8 36
9 32
10 70, 60, 50
11 6
12 38
13 XII
14 3 thousands, 0 hundreds, 7 tens, 6 ones
15 7, 16

### SET 2

1 5469
2 8623
3 7242
4 9213
5 7848
6 4771
7 3043 km
8 4024 km

### SET 3

1 105
2 172
3 145
4 180
5 162
6 368
7 $215
8 216

### SET 4

1 4
2 142
3 70
4 19
5 8
6 Yes
7 $5.95
8 1, 2, 3, 6, 9, 18
9 $24
10 1 L 375 mL
11 Yes
12 0 tens
13 5
14 200 km
15 400 km

### Space

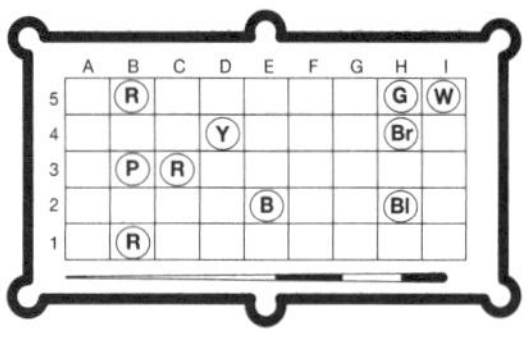

### Space

1 2
2 0
3 6

## UNIT 25 Number and Algebra

**SET 1**

1 49
2 11
3 23
4 7
5 56
6 17
7 63
8 70
9 29
10 28
11 13
12 4
13 259c
14 0 thousands, 3 hundreds, 2 tens, 0 ones
15 45

**SET 2**

1 92
2 120
3 175
4 138
5 96
6 144
7 70
8 114
9 112
10 $175
11 $185

**SET 3**

1 hundreds
2 tens
3 ones
4 thousands
5 hundreds
6 tens of thousands
7 thousands
8 23 296
9 35 413
10 54 086

**SET 4**

1 340
2 3250 g
3 8
4 500 m
5 6 r 4
6 7
7 30 min
8 28
9 44
10 4885
11 143
12 54
13 9654
14 24

### Statistics and Probability

1 OB and MT
2 GG
3 MD
4 100

### Measurement

1 1.2 L
2 3.5 L
3 5.0 L

## UNIT 26 Number and Algebra

**SET 1**

1 14
2 15
3 47
4 10
5 64
6 48
7 70
8 50
9 24
10 8
11 128
12 20
13 379c
14 One hundred and seventy-six
15 24

**SET 2**

1 $81
2 $136
3 $265
4 $162
5 $214
6 $345

**SET 3**

1 15
2 40
3 30
4 25
5 26
6 3
7 4
8 7
9 5
10 6
11 4
12 16
13 20
14 52

**SET 4**

1 $534
2 4250 g
3 1, 3, 5, 15
4 100 g
5 8 r 2
6 10
7 45 min
8 2000
9 39
10 1, 2, 4, 5, 10, 20
11 $1.80
12 36
13 1267
14 $3.80
15 37

### Space

1 b
2 b

### Measurement

1 10:30 am
2 8:05 am
3 5:35 pm
4 10:45 pm

## UNIT 27 Number and Algebra

**SET 1**

1 25
2 2
3 7
4 33
5 24
6 27
7 45
8 36
9 63
10 30
11 10
12 $3.09
13 15
14 0 thousands, 0 hundreds, 2 tens, 7 ones
15 9

**SET 2**

1 8, 16, 24, 32, 40, 48, 56, 64
2 16, 17, 18, 19, 20, 21, 22, 23
3 4, 8, 12, 16, 20, 24, 28, 32
4 18, 24, 30, 36, 42, 48, 54, 60
5 5, 6, 7, 8, 9, 10, 11, 12
Rule: add 4
6 3, 6, 9, 12, 15, 18, 21, 24
Rule: multiply by 3

**SET 3**

1 30 000 + 5000 + 600 + 70 + 4
2 50 000 + 6000 + 900 + 70 + 8
3 60 000 + 8000 + 900 + 20 + 6
4 70 000 + 4000 + 800 + 50 + 4
5 60 000 + 9000 + 200 + 70 + 6
6 60 000 + 5000 + 200 + 70 + 3
7 70 000 + 4000 + 500 + 40 + 1
8 80 000 + 5000 + 900 + 60 + 2

**SET 4**

1 0
2 8
3 30
4 3 L
5 Yes
6 0.21
7 9 thousands
8 $1.05
9 $3
10 1, 2, 3, 5, 6, 10, 15, 30
11 5500 mL
12 $60
13 36 km
14 56 ÷ 4 ÷ 2 = 7 or 56 ÷ 2 ÷ 4 = 7

### Space

### Statistics and Probability

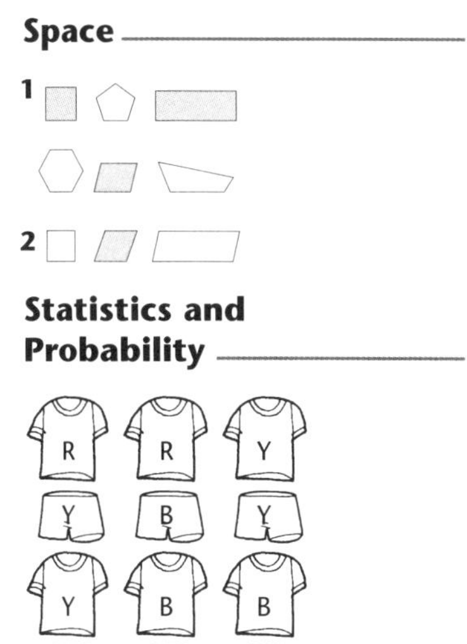

# Answers

## UNIT 28 Number and Algebra

**SET 1**

1 14
2 24
3 5
4 6, 12, 18, 24, 30
5 244
6 9
7 19
8 5
9 365c
10 7
11 7
12 27
13 33
14 $6.59
15 4321
16 $48

**SET 2**

1 $0.40
2 $0.50
3 $0.40
4 $0.50
5 $0.65
6 $0.80
7 $1.50
8 $2.75
9 $3.55
10 $4.60
11 $1.00
12 $1.95

**SET 3**

1 1.26 m
2 1.75 m
3 2.37 m
4 1.58 m
5 3.99 m
6 5.65 m
7 6.07 m
8 1.07, 1.36, 2.07
9 1.32, 2.31, 3.23
10 2.54, 5.24, 5.42

**SET 4**

1 $10.75
2 $25.20
3 56
4 1940
5 17
6 16, 36
7 $2.55
8 85
9 9 r 3
10 6 thousands
11 $297
12 80
13 440
14 5
15 $12

### Statistics and Probability

1 red
2 pink and blue
3 green
4 no

### Number and Algebra

1

| × 3 | 2 | 4 | 6 | 8 | 10 |
|---|---|---|---|---|---|
| | 6 | 12 | 18 | 24 | 30 |

2

| ÷ 5 | 15 | 25 | 35 | 45 | 55 |
|---|---|---|---|---|---|
| | 3 | 5 | 7 | 9 | 11 |

3

| ÷ 4 | 40 | 36 | 32 | 28 | 24 |
|---|---|---|---|---|---|
| | 10 | 9 | 8 | 7 | 6 |

4

| × 8 | 1 | 3 | 5 | 7 | 9 |
|---|---|---|---|---|---|
| | 8 | 24 | 40 | 56 | 72 |

## UNIT 29 Number and Algebra

**SET 1**

1 40c
2 20c
3 $6
4 $10
5 $30
6 $2.50
7 $30
8 $7
9 66
10 29
11 7
12 $9
13 $8 each
14 Nine dollars and fifty-six cents
15 2 half pizzas or 1 pizza

**SET 2**

1 360
2 400
3 100
4 240
5 140
6 200
7 360
8 490
9 350
10 810
11 6
12 9
13 80
14 70
15 $120
16 $250
17 $80

**SET 3**

1 875
2 8.76
3 $47.93
4 $63.47
5 115
6 157
7 $11.50

**SET 4**

1 0
2 0.37
3 350 cm
4 $2.70
5 5
6 1, 3, 7, 21
7 6
8 $12.90
9 Yes
10 Half
11 7 tens
12 0.3
13 10
14 125
15 104 cm
16

| 22 | 8 | 18 |
|---|---|---|
| 12 | 16 | 20 |
| 14 | 24 | 10 |

17 $35.00

### Statistics and Probability

8 9 10 11

### Measurement

1 9 square metres
2 50 square metres

## UNIT 30 Number and Algebra

**SET 1**

1 55c
2 17c
3 $6
4 $3
5 $91
6 $1.50
7 $30
8 $3
9 24
10 50
11 51
12 $2
13 $8 each
14 Sixteen dollars and fifty-eight cents

**SET 2**

1 8
2 8
3 9
4 16
5 7
6 12
7 13
8 13
9 12
10 5
11 11
12 15
13 12
14 20
15 17
16 18
17 22
18 3
19 4
20 6
21 8
22 10
23 12
24 13
25 15

**SET 3**

1 6418
2 2028
3 6327
4 2377
5 4539
6 2122
7 $1174
8 2811

**SET 4**

1 $79
2 4000 mL
3 1, 2, 3, 4, 6, 9, 12, 18, 36
4 250 mL
5 5c each
6 100
7 6:35 am
8 31
9 50
10 1, 2, 3, 4, 6, 8, 12, 24
11 $4.20
12 48
13 Sally

### Space

Hands on

### Measurement

1 12:30
2 $1\frac{1}{2}$ hours (90 min)
3 3:00
4 Price is Right
5 Dr Dill
6 1 hour (60 min)

## UNIT 31 Number and Algebra

**SET 1**

1 11
2 48
3 14
4 56
5 32
6 10
7 30
8 81
9 8
10 $13
11 1350c
12 2 thousands, 7 hundreds, 0 tens, 6 ones
13 22
14 30
15 36

**SET 2**

1 3 r 1
2 5 r 1
3 4 r 2
4 4 r 3
5 5 r 3
6 7 r 1
7 4 r 1
8 3 r 3
9 6
10 8
11 Hands on
12 Hands on

**SET 3**

1 6187
2 7373
3 9468
4 4354
5 1814
6 5919
7 $2212
8 $1190

**SET 4**

1 1250 g
2 14
3 $205
4 1 ten
5 False
6 1378
7 20
8 0.37
9 1.3
10 325 cm
11 No
12 $66
13 $15
14 $55

### Space

### Statistics and Probability

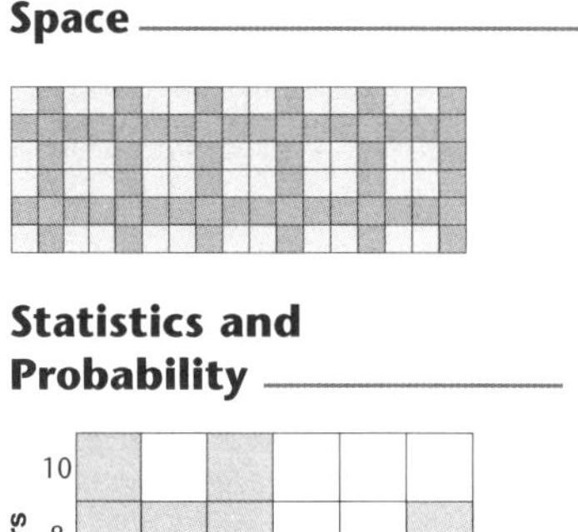


2 Ford and Holden

## UNIT 32 Number and Algebra

**SET 1**

1 30
2 24
3 13 mL
4 30c
5 $1.50
6 2 m 28 cm
7 7 kg
8 64
9 7 m
10 81
11 50c
12 7 m
13 45 mL
14 84
15 16

**SET 2**

1 128
2 258
3 144
4 140
5 180
6 282
7 $84
8 $414
9 5④ × 6 = ③24

**SET 3**

1 True
2 True
3 True
4 True
5 False
6 True
7 False
8 True
9 False

**SET 4**

1 21
2 True
3 5000
4 427 cm
5 8 hundreds
6 0.7
7 0.17
8 $4.50
9 No
10 450 km
11 1, 5, 7, 35
12 0.07
13 10, 12, 6, 12, 22, 20, 12, 4, 36

### Space

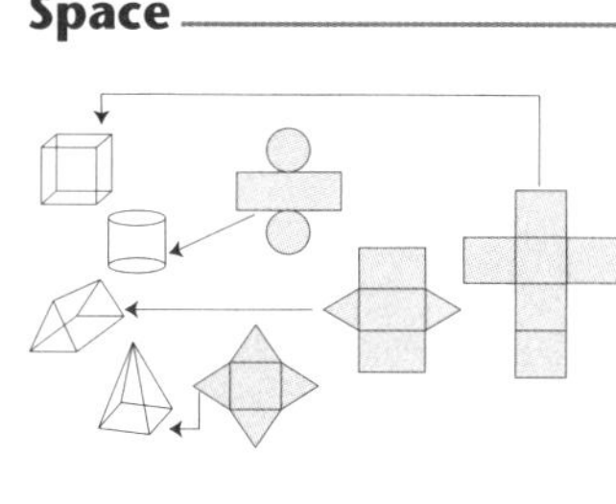

### Measurement

1 29 mm
2 63 mm
3 52 mm
4 14 mm
5 37 mm
6 30 mm
7 80 mm
8 40 mm
9 5 mm
10 55 mm

## UNIT 33 Number and Algebra

**SET 1**

1 18
2 20
3 24
4 11
5 39
6 22
7 27
8 21
9 35
10 31
11 45
12 $24
13 17
14 8 thousand, 7 hundreds, 5 tens, 6 ones
15 $1.80

**SET 2**

1 3
2 5 r 4
3 11
4 13
5 8
6 6
7 8 r 1
8 9
9 9 r 2
10 10 r 1 Correct Bingo card is **a**
11 22 r 2
12 15 r 2
13 13 r 1
14 38 r 1
15 16 r 1
16 17 r 1

**SET 3**

1 $\frac{4}{10}, \frac{5}{10}$
2 $\frac{8}{10}, \frac{10}{10}$ or 1whole
3 $1\frac{4}{10}, 1\frac{6}{10}$
4 $1\frac{2}{10}, 1\frac{5}{10}$
5 $\frac{4}{10}, \frac{1}{10}$
6 $\frac{7}{10}, \frac{5}{10}$

**SET 4**

1 14
2 50
3 16 r 2
4 7
5 3 kg
6 7010
7 1, 2, 4, 8, 16, 32
8 5542
9 12
10 Yes
11 $24.50
12 6300
13 About 500
14 $735

### Space

| | A | B | C | D | E | F | G | H |
|---|---|---|---|---|---|---|---|---|
| 8 | | | | A | | | | |
| 7 | △ | | | | | | X | |
| 6 | | | | | | | | |
| 5 | | R | | | E | | | |
| 4 | | | B | | | | | |
| 3 | | | | G | | | | |
| 2 | | | | | | | | |
| 1 | | | | | | | | O |

### Measurement

1 200 mL
2 100 mL
3 350 mL
4 450 mL

# Answers

## UNIT 34 Number and Algebra

**SET 1**

1 15
2 43
3 25
4 8
5 40
6 4
7 24
8 2
9 24
10 $21
11 1603c
12 36
13

**SET 2**

1 405
2 438
3 580
4 668
5 815
6 894
7 642
8 640
9 992

**SET 3**

1 =
2 ≠
3 =
4 =
5 ≠
6 ≠
7 >
8 =
9 <
10 <

**SET 4**

1 298
2 62
3 171
4 3316
5 6
6 0.8
7 $4.05
8 Four thousand, one hundred and seven
9 144
10 No
11 $3.60
12 1, 2, 4, 5, 10, 20
13 $952
14 4164
15 11
16 5

### Space

Hands on. One example below:

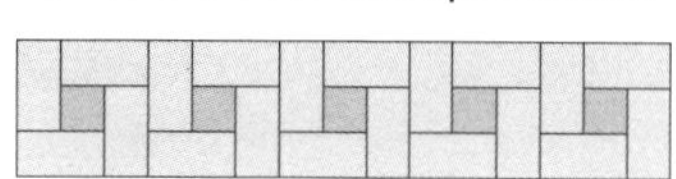

### Statistics and Probability

1 $160
2 $120
3 $20
4 10 June
5 20 June
6 No

## UNIT 35 Number and Algebra

**SET 1**

1 $42
2 $4
3 11
4 $32
5 52
6 51
7 $20
8 1465c
9 56
10 27
11 39
12 6 kg
13 18
14 6
15 70, 60, 30
16 20

**SET 2**

1 ones
2 hundreds
3 tenths
4 hundredths
5 hundreds
6 thousands
7 hundredths
8 tenths
9 thousands
10 tens of thousands
11 thousands

**SET 3**

1 10
2 4
3 55
4 54

**SET 4**

1 37
2 $45.00
3 375 cm
4 Yes
5 $9
6 63
7 $24.50
8 1400
9 20 mL
10 Length 9 m
Width 8 m

### Statistics and Probability

Hands on

### Measurement

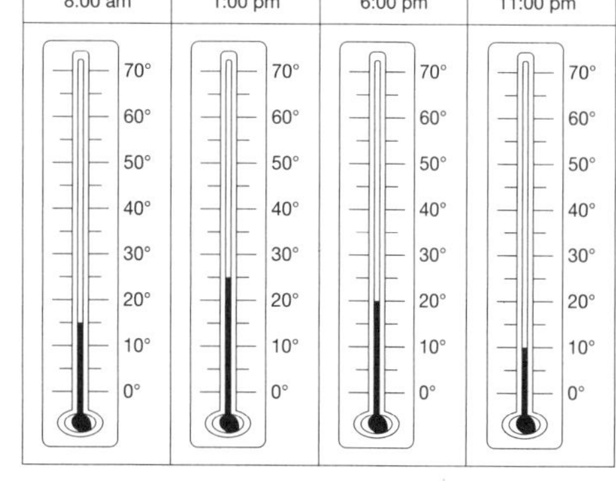